WAVE BY WAVE – FROM CITY STRESS TO OCEANIC FREEDOM

A MEMOIR ABOUT LETTING GO

ANNA ERIKSSON

ABOUT THE BOOK

This is a true story about a couple from Stockholm who sold everything they had, to live and sail around the world. They left a hectic life at the top of their careers, Anna as a professional coach and Arthur as an architect. The soul-searching question Anna brought was: what will happen to us when we take on such a big life change?

The new adventure begins with the summer that turned everything around. You will follow the intense preparation time, from selling the apartment to moving into a new yacht. Anna shares the tough journey of letting go of almost everything. And that is just the beginning of everything they later let go of.

This is a book about sailing the oceans and circumnavigating the world. It is also the story of the inner journey. Taking on a new adventure means being a beginner again. Loving change became a must. Like Odysseus, the story ends with sailing home. This time much wiser after overcoming the many challenges along the way.

Copyright 2025 Anna Eriksson

All rights reserved.

Wave by Wave – From City Stress to Oceanic Freedom. A Memoir About Letting Go

ISBN 978-91-988146-4-4 (e-book)

ISBN 978-91-988146-5-1 (paperback)

ISBN 978-91-988146-6-8 (hard cover)

All photos are taken by the author.

Cover and author photo by Katya@azoresstories

Editor: Lynne Hodgson

No part of this book may be reproduced, stored in a retrieval system, or transmitted in any form electronic, mechanical, photocopying or by any means, without express written permission of the publisher at anna@avalona.se.

*To the curious ones,
who think life is filled with magic
and new possibilities.*

PREFACE

Hello and welcome to my life-changing journey. From a busy life in Stockholm, to living on board a sailing yacht and experiencing the oceanic freedom.

I was on my way back to Europe, on the rough Indian Ocean, when I decided to write this book. I was still alive. My experiences were profound and beyond what I had first imagined. I decided to let the themes that came to me guide me, rather than sharing a chronological narrative in time and place.

From the very beginning, I wrote about the process – in my diary, in blog posts, on social media and so on. Some of these early texts are included to show the process. I felt like a researcher in my own life. Curious about my reactions and responses to all the changes that would quickly come. Everything was new.

The story began seven years ago when my husband and I got the idea to sell everything and sail around the world. It ends when we are back in Las Palmas, Gran Canaria, with a completed circumnavigation. Wave by wave is our motto.

The question I took with me from the very beginning was: what will a journey like this do to me? How would I, a big city woman, like and cope with sailing around the world? What would it be like to be at sea for thirty days in a row? How would it transform me? You'll find my reflections on the inner journey at the end of each chapter after the ✺. Enjoy the adventure!

FOLLOW THE DREAMER IN YOU

"Seize this very minute; What you can do, or dream you can do, begin it; Boldness has genius, power and magic in it." – J. Wolfgang von Goethe

I believe that life is a great gift, and I want to make the most of it. The question is: what is it, then? Goethe also has an answer to that: "As soon as you trust yourself, you know how to live."

What you are about to read is a story about trusting that inner voice. Who told me to jump, to let go of a life I loved. To leave everything behind and go out into the world, on a new adventure. As if life had more to give and show. New things that I couldn't even imagine at this early point.

I invite you to bring your dreamer with you as you read on. The dreamer in you has something wonderful to teach you about life. Come and follow her or him!

My intention with this book is that my story will inspire you to slow down. I have ended each chapter with a question for you to reflect on. Be curious, listen within, and find out more about what your heart is telling you.

PART I
PREPARATIONS

1

THE LIFE IN STOCKHOLM

"To live is the rarest thing in the world. Most people exist, that is all." – Oscar Wilde

I was born in Stockholm. I love this city of islands. Thought I would live there all my life and end my days as ashes in the memorial grave next to my grandmother, Hanna Viola, in the endlessly beautiful forest cemetery, Skogskyrkogården, designed by Gunnar Asplund.

My husband, Arthur, and I lived in a large apartment in Södermalm, the coolest part of Stockholm city, where many young entrepreneurs start small businesses. We also had our own companies; Arthur as an architect and I as an executive coach, with an office downtown. We were in what I call the golden age – healthy, with good finances, children moved out and still healthy parents. Arthur is eleven years older than I; he had just passed the retirement year (67), but that did not bother us. Like many with an artistic profession, he also wanted to work as long as he could. Perfect, because I was in a great place with my own career and business.

Something More?

From time to time, we talked about the future – was there something else, or more, we wanted? For example, my husband thought about buying forest

(68% of Sweden is covered by forest) with a house in northern Sweden. Very well priced at the time.

As a child, I lived for a while in northern Sweden but moved back to Stockholm as soon as I could, so that idea did not get me very excited. And, if he remembered, he wanted a warm climate, not a cold one.

The only thing I could think of was to have an apartment with more light, as I love flowers. We lived on the first floor and my large collection of Flower Maples, in all colours, quickly faded away when we moved in twenty years earlier. Arthur reminded me how bad it was to sell, and then buy a new apartment, from a financial point of view. Too much would disappear in taxes.

And imagine, he continued, how great this apartment is when we get older – we will be able to walk with our walkers to the hospital nearby. He was talking about Södersjukhuset, the nearby hospital where I was born. At the same time, he heard the absurdity of what he had just said and realised that it was not a very exciting future. We were too young to plan our very last days. "Let's create something new and fun", he said.

Some call it the rat race; others liken it to being in a hamster wheel. Either way, the ordinary and predictable life had us hooked. I can't even remember taking the time to think about alternatives.

What does your dream life look like?
Or a wonderful day?

2
THE TURNING POINT

"You are the only one who gets to decide what you will be remembered for." – Taylor Swift

It is Midsummer Eve, and we are at Lökholmen, the home port of the Royal Sailing Club in Stockholm's outer archipelago. The holidays have started, and tonight is the big celebration, the most holy day of the year in Sweden.

The midsummer bar is dressed in flowers, it is a herring lunch with schnapps and song. Continuing with coffee and strawberry cake, a nap before GT at 5 pm and the barbecue. All is well until I take a wrong step and fall backwards down the stairs in the yacht late that night. I immediately feel that something is not right. A rib was broken.

That did not stop us from sailing south, as usual, on Midsummer Day. After a week, we reach Kalmar, a large marina on the southeast coast. I was more or less pain-free and happy to be walking again. I went to the mall nearby. And I fall again. This time, face down on the hard marble floor. A hose was pulled up to my knee level, just as I walked by. I must have angels watching over me because neither my nose nor my teeth broke. I only got scrapes and two blue eyes. Arthur gets mad at me when he realises that the vacation will not be as amazing as we dreamed. The cold and very windy weather does not help either – it is one of the worst summers in a long time.

We Get an Idea

We end up in Listed, a small marina on Bornholm, where we get plenty of time to be with ourselves and talk to each other. We dream together about a future in a warmer climate. I remember an old dream to sail in our yacht to La Gomera in the Canary Islands. The first year Arthur and I met, we did a trip with a friend, Sujatiya, who organised dolphin and whale tours there. That idea makes us think about taking a sabbatical — in five years, or so.

On the way back to Stockholm, we meet the Dutch American couple, Jeroen and Nan. They lived on their houseboat, *Balder*, in the small fishing port of Skillinge in southern Sweden. I share my new idea of maybe living on board for a longer time. Nan takes it seriously and gives me a thorough tour of their yacht and shares how well everything works. Their yacht is large, two floors high, including a bathtub, a library and an office space. She even has a place where she makes hats! I think everything is wonderful.

The Falling

Looking back on my fall, and this was not the first, I can see a pattern. I can see that after each fall, I have made big changes in my life. I have moved, invested in spiritual development, changed careers and started my own company. And after this fall, we would make a radical lifestyle change. The falls are my turning points.

Being hurt helps me be still and reflect on my life. I realise that I need to take care of myself and understand the deeper meaning of what is going on. I do not want to fall again, potentially worse than the previous ones. What do I need to stop doing? What am I not seeing? It is high time to listen deeply to my soul, my inner voice, the greatest lesson ever to learn.

I see cycles all over – it's a good sign that something is alive and in motion. We turn when we reach the bottom, the end of the road or the edge of a pool or beach. We turn when we see no other options but let go of the familiar path and try something completely new.

Over the years, I have thought that those turning points are there all the time, even if they become most obvious when things come to a complete stop.

I haven't always listened so carefully or seen so clearly, so many chances have passed unnoticed. In retrospect, I have noticed that the signals have been there all along, and that they become sharper and sharper when I do not follow them. Existence does everything it can to help me follow my truth.

Now I think that it is part of life to travel over the depths from time to time, in the outer and the inner.

Can you feel whether you're on the way down or up?

3

COMMITMENT TO A BIGGER GAME

"Commitment is an act, not a word."
– Jean Paul Sartre

Back in Stockholm, an intense time awaits. It is obvious that we are ready for a change. Everything goes so fast. I know it is our souls speaking and expressing a longing, free from limitations. We start working again but still live in the yacht at Hundudden, a marina near the city centre. In the evenings, we walk around the beautiful surroundings of Djurgården. Idea after idea flies through us.

Being All-In

The point of no return comes when we realise that we must sell our apartment to afford to buy a new yacht. An ocean-classed, that can take us to places like La Gomera. And, if we shall live on board, why not sail to New Zealand?

I am surprised to hear myself express the bold decision to drop everything and sail to the other side of the world. A dream has to be realistic, otherwise I do not even think about it, I usually argue with myself. That is when it is perfect to get a little help with some fallings from existence.

Commitment

Both Arthur and I know the power of a commitment. It does not take long before we make a final decision to go for this short-lived dream.

Commitment is the ultimate way to consciously say yes to something. It is a free choice and the first important step on a new journey. The wonderful thing about commitment is that it includes a "yes" to being solution-oriented, as challenges and problems are bound to come up. This is how we grow and transform. Commitment comes from the heart, holds trust and gives strength. What is started will be completed. I do not know how, but when the time comes, I/we will make it happen, no matter what.

A New Adventure

Now, when I look in the rearview mirror, I can see how perfect this was for us. The new life would give us a much bigger playground with many new experiences. The changes we had ahead of us were:

- A predictable life, to a life of risk.
- Living in an apartment in Stockholm, to living and sailing on a yacht.
- Costal sailing in the Baltic Sea, to ocean sailing around the world.
- Day sailing, to day and night sailing, also monthly sailing.
- Living with cats, to being without pets. 🐱🐱
- Having many things to a few, mostly spare parts for the yacht.
- Working full-time as a coach, to working part-time.
- Meeting clients in my office, to meeting them on screen.
- A well-planned life, written into a fully booked calendar, to let the rhythm of the sun and intuition guide the day.
- A steady income, to live mainly on savings.
- Swedish, to English.
- Well-known, to the unknown.

We were ready to take the risk. To let go of our old lifestyle, in favour of experiencing something new and hopefully exciting.

The price of the new would be to leave most things behind, including the convenience of living in the city. And soon we would notice that we would also need to separate ourselves from old patterns, habits, and our view of ourselves.

※

When the soul spoke to me, when I finally listened, everything became easy. The soul had other messages for me, than the broken bones that wanted to heal. The soul had a bold plan in the pipeline for us! She knew we had what it would take to make the dream come true.

What is your soul telling you?
What are you committed to?

4

TO BECOME AN OCEAN SAILOR

"Everyone can be a leader." – Werner Erhard

A sailor can be anyone on a yacht. A yacht that you own, rent or are crew or guest on. You can sail solo, to have a large crew. One can sail independently or join one of the many rallies or sailing competitions all over the world. Recreational sailing can take place during holidays or for a much longer time on a lake, sea or even on the ocean. Big differences in both experiences and intentions.

Like many other women, I had never dreamed of sailing around the world. When I met Arthur, I started as a beginner, a passenger. Much later, it was I who said that now is the time. The following events have been crucial to my slow maturation into becoming a partner and ocean sailor.

The Very Best Start

When I got together with Arthur, in 1996, my sailing experience consisted of a day trip in a fast little sailboat, which I thought tilted too much. Arthur was then the proud owner of *Ulla*, a beautiful old classic yacht (A22) in mahogany. He asked if I wanted to come along for a ride. "Yes, thank you", I replied. It was early summer, and time for this year's maiden voyage, 35 NM from Långholmen, to Lökholmen in Stockholm's archipelago. It was sunny with gentle winds. I came barefoot in a long floral dress to honour the day. We stopped halfway at Kalvholmen, outside Vaxholm. He had bought wine, and we had a barbecue. I was in love! With the successful start, Arthur had made sure I wanted to come back on board.

More Adventure

After two years, Arthur buys a bigger yacht with an inboard engine. *Aiolos* (the Greek name for the god of winds), an S30 with a separate aft cabin for his little daughter. For the first time, we sail south of Nynäshamn, all the way to Gotland, the large island in the Baltic Sea. Two years later we round Sweden via Göta Canal. We pass Bornholm, a Danish island in the southern Baltic Sea. I load the boat full of books and read, while Arthur stoically sits at the helm. I call myself a "ruff witch", which usually refers to a woman standing in the galley cooking.

Becoming Co-owner

Ten years later, in 2006, it was time for a decisive decision. Arthur wanted a bigger yacht that could withstand more sea sailing, and he also wanted me to join as a co-owner. We were not married at the time and had neither children nor a house together, so the question felt big, like "Do you take this captain as your husband?" He had sold the *Aiolos* two years before, to save money for a bigger yacht, and sailing friends kept inviting us to sail with them. They understood that Arthur was a captain to the core, and being without a yacht, they also understood, was very difficult for him.

I wanted to make an independent decision – what did I want with my life? Well, two things collided with the summer sailing. Partly, I wanted to go horseback riding in Iceland, and partly, I wanted to do *The Work* – a training about inner wisdom, with Byron Katie. I did those things during our two boatless summers before I agreed. I concluded that boating close to nature was a perfect choice, even for me. We scraped together all our accounts and bought the *Ocean Playmate*, a Wasa 410.

Time to Step Up

Still, Arthur was at the helm, and I was reading. We always sailed south. Arthur wanted to face the heat and attend the Allinge's jazz festival in Bornholm. For years, we talked about sailing to Germany. But it never worked, the southwest wind was always so strong. The coach in me did not like us setting goals that we did not reach. I had to step up, so I took the Coastal Skipper Certificate and bought myself a flotation suit. It would keep me warm as I tacked in the southwest winds with our Wasa yacht, which was completely flushdecked, with no spray hood. And now it was suddenly easy to get both to Germany and further into Denmark. After all, now I am the helmsman. We gain more and more experience in dealing with difficult situations together.

A Patient Husband

Around this time, Arthur plucked up the courage to ask if I would like to live on a houseboat along Stockholm's quays. "Never in my life", I replied. "Forget it!" There were still many other things that felt more important to me. I can only thank Arthur for his infinite patience and wisdom in not pushing me into something I was not ready for. He listened and waited. He also read pilot books, talked boats and sailing dreams with friends and was a member of the Ocean Sailing Club, long before I even noticed.

We had sailor friends, René and Paulien, who emailed us from their circumnavigation of the globe. Our letters always began with "Dear Sailors", and they always replied with "Dear Land Workers".

Developing Together

Arthur and I met at a meditation centre. Over the years, we have participated in many courses in personal and spiritual development. At a time when we had very little money, we invested almost the last of what we had in a course in Bermuda; *Being a Leader* led by the legendary Werner Erhard and his colleagues. We trusted it would be worth it. It was. My business grew, and I got my master's certificate as a coach.

The most important thing I took with me was – a leader, committed to something bigger than herself, will have what it takes to achieve her goal. It means that all our thinking that I am this or that, is not true, just habits and history. On the flight back, we expressed how much we liked the outer islands on both sides of the Atlantic.

A Bigger Adventure

In 2017 came the next decisive step, the turning point that I have already told you about. We got married in the spring and on the summer sailing, the new life-changing idea began to grow rapidly. Now I was the one who realised that if we were going to make this happen, we had to do it now, as Arthur had reached retirement age.

*

For me, the relationship with Arthur has been crucial. Without his listening and safety mindset as a captain, I would never have chosen a sailing life on my own. I am grateful that we are embarking on new adventures together. The most important are:

- We both want to live and sail in the world.
- We both find it fun and meaningful.
- We trust each other and deepen our partnership.

This was about my long maturation as a partner and a becoming sailor. I can see how my energy and awareness increased the more I got involved. What do you think – is it possible to speed up maturation?

Where in life are you maturing?
What do you need to master in your dream life?

5

LETTING GO OF OUR HOME

"Does It Spark Joy?" – Marie Kondo

It was high time to let go of most of our materials – the apartment, our old yacht and stuff from twenty-one years. We took on the roles of massive action queen and king.

Closing the Back Doors

Most sailors keep their home, to return to, during and after the small or large trip. However, we could not afford to both keep the apartment and buy a new yacht. Deep down we felt it was good, and part of the game/process. We also decided to have no storage. Too many had told us that they paid a lot every year before they finally realised that it was more of a burden than something of value. We wanted to take on the new journey fully and go all in.

Choosing What to Release

Since we were going to sell the apartment, we had to start by cleaning and making it nice for photography and later showing.

Then the huge process of going through all our stuff began. I got the perfect help for this big job. During the summer, I read a book by Marie

Kondo from Japan. She has invented a new way of letting things go with respect and without stress. Her focus was to keep only what brought joy.

In short, you go through all your belongings – put them together in five categories (clothes, books, papers, miscellaneous and sentimental things) and then start with the first category and take it piece by piece and *feel* if it brings you joy. If not, thank it for what it has done for you and then let it go, as quickly as possible.

It took time – especially with the books. We lived in a six-room flat with books in almost every room – to be read when we got older. Taken together as a category, they filled our living room floor with tall stacks for weeks. To my surprise and luck, many of them did not bring joy. The same for Arthur who had great books on art, architecture, philosophy and much more.

Sentimental Stuff

There are reasons why sentimental things are left until last, as they are usually the hardest to let go of. We need to train ourselves to let go of other, easier things, first. Usually, I am not very sentimental. I let go, and then I look forward. But this time it was so much of everything.

I had saved my diaries from my whole life. And pictures, of course. I decided to save the very last diaries and let the rest go – they filled a trolley. I came up with a great idea on how to do that.

In Sweden, we have a beautiful tradition on the last day of April every year. We burn old branches from the gardens in large May bonfires. It starts with a long walk where everyone carries torches. When we arrive, we throw in the torches, so that the May bonfire starts burning. Throwing in the torch symbolises the will to let go of the old that no longer serves us. Immediately afterwards, we silently wish for something new. That would be a perfect and respectful way to let go of my journals, I thought. So, I kept them for several months for this particular evening.

It took us at least thirty minutes to walk to the fire with the heavy trolley. Once there, I made myself ready to start throwing my well-kept thoughts on the fire. The guards spotted me immediately and stopped me. "It's forbidden", he told me and explained that it was because of a little more wind than usual. "No", I got so disappointed. You can imagine that the way home was not easy, and of course, it was uphill. I felt like Jesus carrying his cross. The next morning, I threw the diaries straight into the trash. Gone in one minute.

The Art of Surrender

I took my first course in the art of surrender in 1992. It has taken time for this to land, I admit. I received a serious test of my willingness to let go, in 2005 at Byron Katie's training. One morning we would leave the most valuable thing we had with us. I did not have much, but the dearest was a black waistcoat, which I left with a heavy heart on the table. I felt so miserable about it, so the assistant checked in with me. No wonder this brilliant life-changing idea did not come up then. I was far from ready, as it would mean letting go of much, much more than an article of clothing.

But even now it was hard to let go, especially about the books. A wise friend, Prajna, encouraged me and said that I have everything I need with me anyway. She was right, I have not missed anything.

Separation

Every single choice is a moment when I say "yes" or "no." With "no" comes a separation, a goodbye, and a thank-you. Old memories with the accompanying emotions pass at the same speed. You can probably imagine what a roller coaster it was some days. Without a break comes the next round of decisions – to sell and if so where, to give away and if so to whom, or to simply throw it away. For us, this was a constant process for eight months. Each decision also made our new life clearer and more tangible. It was very satisfying to know that the relatively few things we kept also brought joy.

All spiritual practices involve letting go of material things. Yet it is something most of us hold onto for as long as we can. Although it felt very heavy during the process of letting go of most of our possessions, it was surprisingly easy afterward. This was just the beginning. I would soon discover how much more there was to let go of.

How are you with things – do you find it easy or difficult to let go of them?

6
MONEY

"Since money is energy, our financial affairs tend to reflect how our life energy is moving. When your creative energy is flowing freely, often your finances are as well. If your energy is blocked, your money does too." – Shakti Gawain

How much money would we need as a buffer to dare to leave the workforce and invest in a new kind of life? Some sailors have calculated the annual costs, so we had a rough idea. We knew that my husband received a small pension, and I hoped to continue coaching. Since we were not financially independent, we wanted to get as much as possible for our possessions.

We were soon forced to realise the low value of our dear things. If I ever create a home again, I know that I can get very good quality furniture at auctions and second-hand shops. In Sweden, that means selling at Bukowski's, Metropol and Blocket. For example, we had beautiful real rugs that we got almost nothing for, even though we had bought them expensively a few years earlier.

However, these losses were nothing compared to the apartment. Prices had gone up every year since we bought the apartment. When we came home with the idea to sell in a year or so, the whole market started to drop. We decided to sell as fast as possible, but we kept the price too high. It took longer to sell, and of course, we had to adapt to the market in the end.

The same thing happened to our previous yacht. Although it was a very good one, she was old, and newer ones were more sought after. We almost gave it away for the price of the new engine we had recently installed. Two weeks before we left.

Our buffer to live was rapidly diminishing, but we knew it was too late to change our plan. The finances were just another challenge that we had to trust that we could solve later.

Spending Instead of Earning

After these sales, which at first felt like losses, we needed to invest heavily. First in a new yacht and then in upgrades for the round-the-world sailing. Fortunately, we were not aware of how much it would cost at this early stage. Later, we understood that the first year is always the most expensive. After five years, we know that living on a yacht is not cheap. The yacht consumes as much or more than our living expenses.

Money Is Energy

> "Energy cannot be created or destroyed, it can only be changed from one form to another." – Albert Einstein

In hindsight, I can see that this first period of selling and investing was a good lesson in using money as an exchange. At least for my mind, which prefers to save instead of spend. It was like a test of whether we were ready for big changes or not.

I can also see that we both had the energy to make our dream come true. We ran our own companies, which is a big responsibility to create both value and revenue. And we were the ones who had generated that we were living in an apartment that we could sell and get out of debt at this time in our lives. In Stockholm, many apartment buildings were sold and became condominiums. When it was our turn, Arthur was on the board and knocked on everyone's door to get a "yes" to buy. A third of the apartments were collective housing, and most of these tenants were against houses becoming privately owned. In the final vote, the "yes" side won with a majority of one vote. This purchase and later sale gave us a new yacht and allowed us to try a whole new way of life. I am very grateful.

We do not get happier just by being richer. I have met CEOs with tons of money who are mostly worried. And I have met others with apparently little money in their accounts, who spread joy and who love their lives. How we see and think about ourselves matters.

The second thing is how much we use the money – do we dare to let them go and let them flow and change form? I have noticed that when I give myself something I really want, it always takes me forward.

What is money for you?
What are you investing in?
Do you invest in your own development?

7

TO HAVE AN INTENTION

"Life isn't about finding yourself; it's about creating yourself. So live the life you imagined." – Henry David Thoreau

I love the word intention. Out at sea, all boats and the Sea Rescue use VHF for communication between them. We start by saying the name of the boat we want to call three times, and then we say our boat's name. The next question is – what is your intention? To answer, I need to know what I am doing out here, where I am going and what I want when I make contact.

The distinction, of intention, is useful in many contexts. Arthur and I expressed our intention for this project like this:

- To raise our energy and consciousness
- to see and share the light and beauty of the world
- to explore new parts of the world – inside and outside ourselves
- to have fun
- to explore being
- to expand
- to feel, experience and express existence, the essence
- inspire others to live a life they love
- to trust and express our creativity.

There are many questions we wanted to explore. The most important are:

- What will happen to us during this journey?
- What is it like to live with very few material things?
- What will our trust and creativity bring?
- How does it work to make big life changes, at our age?
- What is it like to be a world/ocean citizen?

Plan A

We decided to have New Zealand as our destination – as far away as possible when we have the most energy and money. There, we said, we will evaluate whether we want to continue the sailing life or not.

For the journey, we set milestones in time (months), nautical miles and destinations.

I have noticed that there is a very high chance that what I declare out loud, and/or in writing, also reinforces it happening. It is of course risky to let others know what I want from the beginning, but it also works the other way around. It helps me keep my word and maintain my integrity.

Do you have outspoken intentions for your day/life?

8
COMPLETING THE WORK

"You are complete. Completion has taken place in you eons ago. Accept yourself. Enjoy yourself. This time will never be again, so fulfill the moment with your own self-joy." – Frederick Lenz

It's one thing to commit to a new lifestyle. The other is to be in integrity with what I was already committed to, such as work, and to complete it properly. My clients sign up on an annual basis, so I knew I had to stay in Stockholm until next summer.

This was tough for me. I had collaborations with clients who counted on me for a long time. How would I tell them and still maintain trust? Would they believe me with this sudden change? I mean, even I thought until recently that I would be working as a coach in Stockholm for many more years.

I was soon faced with the dilemma of signing new clients, even though I was honest about my plans. My husband solved this by being a senior consultant to a colleague. It was a perfect role for Arthur, who could share his wisdom with the many young architects.

As if that was not enough, the entire building where I rented my office was to be completely renovated. Everyone had to move. It is not very difficult to find an office in Stockholm, but I loved being in that place, in Lästmakargatan, in the heart of Stockholm. It worked out, of course, and I loved the new place in Barnhusgatan too.

On Friday the 17th of May, I had my last client meeting. I left the keys to the office and walked out into my new life. I felt empty and the future uncertain.

I knew that completion was important because I needed as much energy as possible for our new project. Completing the work meant saying goodbye to clients and colleagues, but also my physical office, furniture and even more books. Closing those doors too. Having a deadline made it clearer.

The hardest things to deal with were areas where I was blind/unaware. My books, for example. I lived in my dream world that someone would buy them, even though the antiquarian had refused them. It took energy to be indecisive and think about them, instead of making a quick decision and getting rid of them.

Grief is natural when we leave something or someone behind. Ceremonies, like a funeral followed by a gathering, are a way to deal with endings. Doing something beautiful and memorable where we support each other through all the emotions and memories that may surface. We know that an era has ended. When we leave a job, we usually get flowers and some kind words. As a self-employed person, I had to deal with it on my own. Looking back, I should have asked friends to help me.

How do you envision completing your working life?
Or if you're not there yet in age — completing the work you have now?

9

THE CALENDAR

"Creativity involves breaking out of established patterns in order to look at things in a different way." – Edward de Bono

Another change occurred this spring – small, but significant and perfect. I was one of those who carried around a large Filofax, an A5 in brown leather. I loved it, especially the visual overview. As part of the process of letting go of all our stuff, I decided not to buy a new calendar as the new year approached. Now was a good time to go all digital.

Bullet Journal

Interestingly, a new type of calendar appeared after we left Stockholm. It is called Bullet Journal, or BuJo. It is a way to make your own calendar with a future log, monthly, weekly and daily log. Besides the future log, you create the rest of the calendar as you go. In that way, you can create and use the space you need for tasks and reflections. Different depending on what is going on.

I started very simply and noticed that it suited my new life very well. Especially for capturing ideas and reflections. I still use this system, and I can see that it reflects the creative side of me. Every day starts with a blank page.

This was just the beginning of controlling less and trusting more. A white page is an invitation for something new.

Are you setting aside time for reflection and journaling?

10

BUYING A NEW YACHT

"Let your home be your mast and not your anchor." – Kahlil Gibran

We had the most important decision before us. We needed a yacht built for safe sailing on the oceans. Comfort and space were also important, as we would be living on board full-time. This and next chapter will be very boat specific. Welcome to my new world – with our new yacht being our home!

Already in 2008, we started making a list for our new dream yacht – at that time still for sailing in the Baltic Sea. No. 1 on the list of 32 qualities was beautiful (Anna), and No. 2 was seaworthy (Arthur). But the main thing was a doghouse, to protect us from the prevailing southwest winds in the Baltic Sea.

Choosing a Brand

The only brands my husband knew were the Dutch Bestewind and Swedish Hallberg-Rassy (HR). Arthur wanted a yacht that was still in production, so we could be sure of getting spare parts for a long time. We visited HR to see how it felt, especially for me who loves big spaces. To my surprise, I felt it was perfect. She was forty-eight feet, around fifteen meters long. I have always loved their aft cabins with a large bed. I also saw a place where I could sit and write and where I could put my printer.

One day we met Stefan and Pia, who had returned from the Caribbean.

We shared our new dream. They knew exactly what was important and a must on a yacht for full-time living and sailing. A generator, a water maker, a freezer, a washing machine and so on. I wrote everything down on a piece of paper. They suggested we look at Amel yachts. They had met many along the way and the owners always said they love their yachts.

That same evening, Arthur Googled Amel. It was a French built ketch with two masts. It looked a little different. I got immediately interested.

Plan A to C

As soon as we had sold our apartment, we planned to buy a new yacht abroad. Depending on when in time this would happen, we made plans.

a) Our first wish was to buy a yacht in time to be able to sail her to Stockholm during our upcoming summer vacation.

b) In the Nordic countries, autumn is a stormy time and in winter there is ice. If we were to get a boat home at all during this time, we needed to hire a skipper.

c) And finally, if the sale and purchase took a lot of time, we thought we would ship our things down to where the new boat was. Much more complicated, of course. And we would also need to find temporary accommodation in Stockholm.

Luckily, we found our yacht at the last minute to fulfil plan A.

Buying Abroad

In the middle of June, two weeks after we had sold the apartment, Arthur found an Amel for sale in Toulon, on the French Riviera. They are popular and not so easy to find. We flew down to Nice. The broker, Michel Carpentier, met with us and asked if we had seen any Amel before. "No, we haven't", we replied. I saw how he got nervous, and he asked us, "Do you know that the floor is blue?" As if it was of great importance. "No", we did not know about that. Our main concern was whether the aft cabin was long enough for both of us. This later redline model was built for the US market, and the bed size was one of several points that differed from the earlier model.

We met the French owner on board *Vista*. His story was sad, this was his dream, but his wife never liked the sailing life, so we think he faced an ultimatum. Her absence was very visible, and the interior completely lacked atmosphere. They were the second owners, and neither of them had sailed outside the Mediterranean.

When we were done with looking at the yacht, and were alone with the broker, Michael asked me, "What do you think of her?" He saw that I was not very happy. And it was true. I did not feel her beauty. It took some time and talking before I got over the first impression and understood that just that, the surface, was the easy part to change. As Arthur was super happy with the yacht itself, we agreed and made a 5% down payment.

Boat Survey

We requested a boat survey before making our final decision. The best inspector of Amel yachts, Olivier Bateaux, could not come at such short notice. We had to accept a local inspector, to be able to sail back to Stockholm on our holiday.

Ten days later, Arthur flew down on a one-way ticket and took part in the survey. The result showed that *Vista* was in good shape. The batteries that boiled when we were there had been replaced, and a few other things were also fixed.

Afterwards, we found many other things that needed to be upgraded or replaced. A sign, it was time for *Vista* to get new owners who took care of her.

The Purchase and Handover

When the survey was done, Arthur called me, and we decided to buy *Vista* – the first and only Amel we had seen up to that point. I arranged with the bank so they could make a large transfer possible. Then I also flew down on a one-way ticket. We slept our first night in *Vista*, and the next morning, we met with the owner and broker for the signing. The owner brought croissants and a bottle of champagne for us to celebrate with. He himself was close to crying. We felt that this change of ownership was the best for the yacht.

Then the broker took us out on a sea tour of the *Vista* and showed us the special pole system and other typical Amel details. He also taught us to dock stern to – the Mediterranean way. After a few intensive days of cleaning, AIS installation, ordering insurance and food bunkering, we were ready to go.

Take Vista Back to Stockholm

We sailed her back to Stockholm on our six-week vacation, making only nine stops on the entire trip. For the first time ever, we sailed day and night. We took three-hour shifts at night. I was afraid of not getting enough sleep. From the very beginning, I learned to rest as soon as I got tired, even during the day.

We were afraid of all the new waters ahead. First Gibraltar – calls to VHF were intense with Mayday – people fleeing and wanting to be rescued. The next challenge was the coast of Portugal, where northerly winds usually blow. We got lucky again and arrived just as the month's weather window opened. Then Biscay. And the English Channel. And the North Sea. Despite this fear, Arthur's confidence grew. He remembered he knew how to make a landfall. We had a backup plan with a friend who was ready to come and sail with us. Pretty soon, we both felt like we could make it on our own – a good first test.

Marina for the Winter

Finding an open winter marina in Stockholm is not the easiest. The archipelago freezes into ice. The majority of all boats are ashore between October and the end of April. In Stockholm's inner city, there are three marinas to choose from. They are all in Djurgården – a beautiful park-like recreation area with museums, art galleries, Tivoli and an outdoor zoo. Arthur had already arranged for us to have a place at Wasa Marina – the closest one to town and the only one that had the services we needed.

The only problem was that when we returned with the *Vista*, the owner had forgotten about it. We knew this marina only had a few spaces for winter boats. The waiting list was long and most people, regardless of where they were on the list, were turned down by Leif. Rumour had it that one in hundred got a yes. Standing in his small office, we understood that it was a critical moment. We did our best to share our story and great need for a place. Luckily, it did not take too long before he said he would fix us a place. It turned out that the people he wanted in his marina were like us, on the way out into the world. I was super happy about this.

Naming a Yacht

I had dreamed of having a yacht called *Gratitude*. Arthur had agreed to change the name according to all the rules of the art when we got home. It was near Gibraltar when we changed our minds about renaming *Vista*. On the VHF, we heard *Sweet Spot* calling for help many times without a response. I was not sure if it was a man or a woman speaking, but the voice was sweet. After a long while, *Depression* answered in a male baritone voice. It sounded so strange. I felt that a neutral and simple name, like *Vista*, is much better. Especially in emergencies. I can express my gratitude in many other ways.

A Very Good Choice

Now, with five years of experience sailing *Vista*, we are more than satisfied. Like many other Amel owners, we love our yacht too. She handles very well in the sea. The centre cockpit with sails on electric winches makes her very safe. Thanks to the deck of fibreglass, and not teak, she is completely watertight. The two masts were new to us. The smaller Mizzen Sail has proven to be a good stabiliser, which also gives a little extra speed. And we love the blue floor.

Living on a yacht is a condensed way of life. In a small space, the most vital things for both living and sailing must have their space. This simplicity appealed to me.

The special thing about living on a yacht, compared to a house or apartment, is that we are always moving, even in a marina. We live in, or on, the element of water, rather than the earth.

What qualities are important in your home?
Where is your ideal location?

11

THE MANY FIRST PREPARATIONS

"God is in the details." – Ludwig Mies van der Rohe

We came back with *Vista* at the end of August, just before we were supposed to start working again. Now we had less than a year to get everything ready – to transfer our living and ourselves to the new lifestyle.

Many have wondered what has been the hardest. This spring in Stockholm was the toughest for me. Not only had we let go of most of our belongings, which brought sadness, we also had so much to think about and organise. Plus, all the new things to learn.

It will get easier, I promise. Stay with me and notice your own feelings in the meantime.

Downsizing

We had September to move our last things down to the yacht. Our exit from the apartment was well planned – people came and collected things day after day. The last night, we slept on the bottom of the bed before we carried it out to the garbage container.

The maximum weight to maintain balance and safety in the yacht was 3000 kilos. That may sound like a lot, but that weight includes 1600 kilos of water and diesel. Arthur weighed everything that was going to come on board. In the end, we had to let go of even more stuff.

It was a special moment to hand in the keys – twenty-one years of living in this apartment ended and a new life began.

Living Onboard Wintertime

Our French lady was not built to be in freezing water and be covered in snow. She came with air conditioning, not a heater. Very soon we were putting agitators around the yacht, to keep the water open right around her. And then we covered the deck with insulation material and tarpaulins. We bought three large water elements and a dehumidifier to keep indoors – with them, we had 20° all the time.

The heat inside combined with the cold outside produced a lot of condensation. We poured out five litres of water/day and sponged out water from the bottom of the wardrobes and in the spaces under the floor. Even worse, it also produced mould. I quickly learned to treat it with vinegar.

We will never winter in the Baltic again. *Vista* is built for warm water. The cockpit and the external environment are part of the living area. For this first winter it worked, thanks to the beautiful surroundings where we took our evening walks.

Preparing Vista for Ocean Sailing

Sailing on the oceans requires much more focus on durability, spare parts and safety compared to coastal sailing.

We had all the machinery and rigging checked.

Arthur built a third fridge that could be converted into a freezer. He changed the gas bottles to propane – better in cold climates.

And we bought a lot of things:

- a new mattress to cover the entire aft cabin
- solar panels
- an extra autopilot
- a new chart plotter and charts for the whole world
- a new heavier 40 kilo Rocna anchor
- a longer 100-meters anchor chain

- new ropes for docking
- lots of spare parts
- safety items: Iridium Go, EPIRB, distress signals, safety line, extra AIS, AIS mob, life jackets, fire extinguishers, mosquito nets
- flags for the whole world
- diving equipment
- fishing gear
- electric stove, thermos and lots of storage jars
- a large ship's medicine cabinet
- a new dinghy and an electric Torqeedo outboard motor

There was also a lot of administration. A boat over fifteen meters, must be in the ship register and have a nationality certificate from the Swedish Transport Agency. We ordered a new call sign and MMSI number that we had programmed into the safety equipment.

We changed the labelling on *Vista*, removing the TL for Toulon and replacing it with Stockholm. A fun thing was making our boat stamp.

We ordered international insurance for both the yacht and us.

We joined several international yachting organisations. The English Ocean Cruising Club (OCC) proven to be the most active around the world. We were already members of the Swedish Ocean Cruising Club, OSK. Arthur has been a long-time member of the Royal Swedish Sailing Society, KSSS.

Preparing Us

Arthur and I took several trainings at the Swedish Ocean Cruising Club in Stockholm.

- SSB radio to see if we should invest in that type of equipment. It was excellent training, but we felt that the short-wave radio had had its day and that Iridium, and later Star Link, had taken over.
- Monthly group meetings with other sailors. About forty yachts were also considering leaving Sweden in the coming season.
- Sextant, how to find out our position using the stars.
- I did the longer course for ocean sailing, equal to Ocean Master. Arthur already had this.
- I also got my own VHF certificate.
- Arthur bought pilot books and learned more about the routes.

During the spring, we got certificates in diving.

⁂

This was a complex year with many mixed feelings, especially for me. Arthur was happy as senior consultant and focused on getting *Vista* ready for our new adventure. The vision of letting go and sail south and explore the world kept us going.

In hindsight, I can see how perfect everything was. This time taught me a lot about both letting go and the importance of taking care of details. Arthur reminded me of the architect van der Rohe quote above.

It was also a lesson in being with my impatience. I wanted everything to go more easily, faster or differently – exactly my sore spots to transform. We were tested – were we really ready for our new dream?

How would you approach a big change like this?

12

WAVE BY WAVE

"The journey of a thousand miles begins with a single step." – Lao Tzu

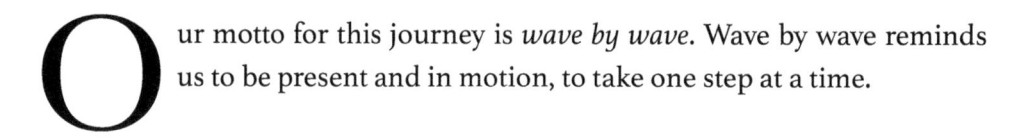

Our motto for this journey is *wave by wave*. Wave by wave reminds us to be present and in motion, to take one step at a time.

Wave by Wave in My Company

Wave by wave was invented when I started Avalona, my company. I was struggling to find new CEO clients. My coach, Liz, asked me when I felt in flow. What came to me was when I was at the helm and steering. A key aspect of running Avalona was including and letting my inner woman be with me. So, this new character became a water goddess, steering her life and taking it wave by wave. It was perfect and helped me be present in the here and now, no matter what or who was in front of me.

Wave by Wave as We Sail

We decided to use the same motto for our sailing and named our blog wavebywave.se. Now, we have literally sailed around the world, wave by wave.

Wave by wave does not mean taking it easy and trusting everything. It means preparing by listening to each other's ideas, thoughts and priorities. We talk through what to expect and later how it went. What can we learn and do differently next time? We are open, curious and do our best to be with what is.

The motto has served us very well, especially when it comes to making decisions. What next step is available?

Lesson learned: if you can move in the direction you want – go! A small next step is enough.

I am so happy we came up with this motto. It helps me not to get overwhelmed and to be present to what is.

If you have a motto – what is it? Or what would it be?

PART II
LEAVING STOCKHOLM

13

LETTING GO OF SWEDEN

"Holding on is believing that there's only a past; letting go is knowing that there's a future." – Daphne Rose Kingma

It was a grey Saturday morning, May 25, that we let loose for good. Exactly on the day according to plan A. Some friends waved us off on our way out through the archipelago. It was a strange feeling to leave Stockholm behind us, not knowing when, or if, we would ever return.

Family and Friends

I don't want to say that we left family and friends behind. I hope they will be there wherever I am in the world. I remember with warmth the friends and our mothers who came to visit us in Wasa Marina. Arthur's father passed away in January of that year; it was not unexpected, so in a way it was a good ending. And my father left the life long time ago.

We also said goodbye to our fourth cat, Kovo. He turned fifteen, and we knew he didn't like boating. He rests in peace with his mother Snufsan, aunt Donna Valencia and old mother Rosa at the animal cemetery on Djurgården.

Leaving My Hometown

Stockholm is the city of my birth. I will love this city forever. I have walked and cycled around a lot, especially in Södermalm where we lived. On the short trip into the city centre, I often stopped with my bike at Slussen. It is a large lock where the outer sea, Salt Lake, meets the inner lake Mälaren. The view is magnificent and includes the Old Town and Djurgården.

We left our favourite shop: ICA Aptiten which sells excellent cheese and fish. And Hötorget, the outdoor market, with flowers and vegetables, and the indoor market with fresh lamb and olive oil. I left my skilled cobbler. And Andy, my hairdresser of fifteen years who retired at the same time I left the city. I left a lot of nice lunch places in the city, like Mamma's and Linda's.

Leaving Step by Step

We left the country step by step. Our first stop was Sandhamn. The place in the outer archipelago where we spent all the weekends before and after the holidays. We had one last dinner with my brother and his family. The next step was Kalmar to meet the Hugossons, the couple who led us to an Amel.

We planned to continue directly to Kiel, but it was as if the air was running out of me, a little outside Kalmar. We had a strong south-westerly wind, the normal for this season. I was simply too tired to continue, so we sought shelter in a small village, Ekenäs, where we had never been before. I saw it as my chance to say goodbye to the Swedish flora and light.

A few days later we left Sweden and the very well-known Baltic Sea. This time we were better prepared. It was easy to pass through the Kiel Canal – the umbilical cord to the North Sea and the new unknown. Now we were part of the small group, a handful, of Swedes who every year set out to sail around the world.

Have you ever thought about living abroad? If so, where to and why that place? What do you think would be the hardest for you to leave behind?

14

BACK TO THE CRADLE

"The privilege of a lifetime is to become who you truly are." – Carl Gustav Jung

Our first goal is to sail down to La Rochelle, where they build all Amel yachts. We have a slot time at the beginning of July.

La Rochelle, at the inner end of the Bay of Biscay, is a beautiful old town with a huge marina. They have a long tradition of building yachts here. Besides Amel, they build charter yachts like Dufour, RM, Beneteau, Jeanneau, Sun Odyssey and the catamaran Lagoon. Outside the market hall, they serve fresh oysters for lunch, for only €10/plate!

Amel has two pontoons in the marina, no. 50-51, for new Amel yachts and others, like us, that come for some kind of service or are newly purchased or for sale. At the entrance, Monsieur Amel's last yacht is docked. It is now owned by Jean-Jacques, the former marketing manager and later CEO and chairman.

Vista is quickly taken care of. Already on our second day, we are up on land in 35° summer heat. Sebastian services the bow thruster and other Amel-specific parts. The team around Amel are all very professional, skilled and fast. They understand the vulnerability of being out on the seas and are keen to fix everything, so that we can be self-sufficient later.

We visit the Amel factory and meet the legendary Maud, who takes care of all orders for spare parts. Terry, the new marketing manager, takes us on a tour, and we get to see how well Amel builds their yachts. The most significant and unique to Amel is the invisible. The yacht remains in its mould for five months, throughout the process – from the manufacture of the hull to the insertion of the interior and engine. This makes the construction very durable. Only after this do they move the entire yacht to a pool, to test the engine and electronics. The final step is to launch the new yacht, at the so-called Amel baby pontoon, in the marina.

The common way, used for the large charter industry, is to move the hull from its mould as quickly as possible. That is a cost-effective way to make room for the next hull. The risk with that move is that the hull may suffer settlement damage during transport. The alignment process, which must be dealt with before the internals and the engine are put in place, can never provide 100% precision. Monsieur Amel never wanted to compromise on this. He wanted it to be perfect all the way through.

It is a basic human need to want to belong. More and more people move from their birthplace on earth. I thought I should live and die in Stockholm, and now I see myself as a world citizen.

Here I tell you about Vista's cradle, but how do we remember our own cradle?

15

IN WONDER

"Don't call it uncertainty – call it wonder. Don't call it insecurity – call it freedom." – Osho

I remember how I felt as a child in our first months – in a sense of wonder and curiosity for everything new. I wrote the following in La Rochelle, a month into our new life.

We are off on a new adventure. I am still surprised that, as a Stockholmer and an academic, I am on a yacht that intends to sail to the other side of the world. As a child, it was completely abstract. I did not know anyone who lived or had been to the other side of the world. Flying or sailing was not "in the cards", much less taking a trip around the world after high school, which is common these days.

A month ago, we left Wasa Marina on Djurgården. Now we have arrived in La Rochelle in France, where the shipyard for our Amel yacht is located.

Sometime next autumn, we will reach New Zealand, if everything goes as we have planned right now. It's at least 10,000 nautical miles away, via the Canary Islands, the Panama Canal and the entire gigantic Pacific Ocean. The longest distance we've done so far was 3,000 nautical miles, from Toulon in the Mediterranean to Stockholm, when we brought *Vista* home. A long-distance and big stretch compared to the around 1,200 NM during our holidays.

My first sailing trip with Arthur was about 35 nautical miles, which we took in two days so that it would not be too far for a beginner like me. Things have changed.

Our new adventure is something completely different from a slightly longer holiday. We completely changed our lifestyle. From a big city to the world, from an apartment to a yacht, from work to sailing, from the archipelago to the sea. We have so much to learn! It is like stepping into a new world. We are starting from scratch and hope that our trust and creativity will carry us all the way.

I feel like I have thousands of questions all the time. I feel like a child who wonders about almost everything. How will I experience everything I encounter? I hope that I will become wiser when I find myself in completely new contexts.

We are in a phase where we adapt ourselves and the yacht to function optimally. Last fall, we moved into the yacht with a fraction of our household goods. A fraction is still a lot in a boat, and we are now thinking about what is most important to keep. Last winter, we only lived on the yacht, which meant that we had a lot of stuff in front. Now, underway, everything must be sea-stowed and withstand constant movement. The Riedel glasses are well-packed, as are the porcelain bowls. Plastic is perfect on board. It was a big project to find good storage that keeps both moisture and small animals out. We have not come across them yet, but it seems only a matter of time.

As for our approach, it is about finding new routines for life on board and new ways of communicating with each other. Now we are together all the time and have to cope with our joint project and the challenges and opportunities that naturally come with it. When we sail, most of the energy goes into sailing the yacht, navigating, checking the weather, cooking, eating and resting/sleeping. Sometimes when it is very quiet, reading or writing works, like now. We have introduced morning talks, and in the evening, we share what we are grateful for (a lot) and what we have learned today (a lot).

Yes, this is how my thoughts go right now on board the *Vista*, at the very beginning of a new journey.

I love the feeling of being in wonder. The first impression of experiencing something new, is always unique. Moments to cherish, because they will always disappear. The first time you meet someone, the first day at a new job, or the first day in a new place. Or a new idea. Anything that is new. It gets me feeling alive.

Do you remember the last time you felt in wonder?

16

RESTING IN THE RÍAS

"All of humanity's problems stem from man's inability to sit quietly in a room alone." – Blaise Pascal

"I have thirteen squirrels in my head reminding me of everything I have to do.", says my neighbour on the other side of the pontoon in La Rochelle. The French couple, in middle age, are new owners (only ten days) of an Amel. She lets me know that they have three houses, where everything seems to happen. I sense her joy, excitement, and stress.

I so recognise myself in her. Even though we have sold our only apartment, and do not even have storage, I hear the chatter from my mind too. Even my/our list is still very long. A perfect time to practice some meditation, witnessing what is, without acting on it. It was so easy as a child – I lay on the meadow and watched the clouds pass by.

I thought this stress would be over soon. Like when I am on vacation, and the new energy comes back after a few weeks. This time it took years. Perhaps the fatigue dared to appear because we were free in a new way. I understood that we needed time to integrate. Not only the new lifestyle but also the new impressions we got along the way.

After La Rochelle, we had planned to sail in the Spanish rías. It was a wise decision. We gave ourselves time to rest and be slow, to rock in the hammock, to read the thick books I never finished at home. In the rías, I wrote the text below.

Rest is important, we notice.

Both to land in all the new, and to get deeper into ourselves.

Incredibly nice to have time to be, think, read and talk deeply.

Being gets more space.

On board, it is a different pace.

Everything takes longer than on land.

We pack up and down the dinghy, or portable bikes.

We stow and stow again.

We polish and clean.

And the greatest luxury – we wash on board and dry in the sun and wind. That smell!

We have about four months before it is time to cross the Atlantic. I have thought a lot about whether we should bring someone along on the passage. We concluded that we must manage it our selves, as we will continue directly to the Pacific via Panama.

We go to the other side of the world first, while we have the most power.

Much on board is about preparation.

Having enough spare parts and food with us.

We have bought cockroach traps – they are down south in Cascais (Portugal), and here too, we have been warned.

We are thinking about which ten dishes to duplicate and cook when we sail over to the Caribbean.

We learn how to cook a one-pot meal in the pressure cooker.

We cook lentils and beans – it is a staple that is easy to take with us.

We have baked the first bread on board. It did not turn out so well – the dough did not rise as it should.

And of course, we tried all the local delicacies. Here, the water is rich in different kinds of fish, mussels and shellfish.

And I still wonder if there is food to buy in the Pacific... there must be! Canned food is not so much fun, for food lovers like us.

Arthur reads which ports we can go to, as the anchoring opportunities will decrease in the future. The prices differ a lot. A guest port night can cost like a hotel night – it works for a vacation, but not for a lifetime.

And we check the weather – many times a day.

Yes, we learn to live a new life with new structures, routines and rules of the game. All to make life on board work in the very best way.

It is fun and we have a good time.

It is still unimaginable that we are here in the heat and sun and will continue forward and break new ground.

Solar winds from us in Cangas, Spain (August 21, 2019)

It was a very good start to rest. The more relaxed I become, the more I got in touch with my true self.

What is your way to handle stress and the mind chatter?

17

WHAT YOU GIVE COMES BACK

"When I let go of what I am, I become what I might be. When I let go of what I have, I receive what I need." – Tao Te Ching

We are at anchor off Getxo, on the northern Spanish coast. The Guggenheim Museum is nearby, in Bilbao, and of course we want to take a look. We take the dinghy ashore, and then the local train and a short walk. The museum is impressive with a giant dog sculpture, filled with flowers, outside the entrance.

To our delight, there is an exhibition of Anselm Kiefer on the top floor. Arthur had his books in front of him as he worked, for great spirit. It was not easy for him to sell those dear books, but he did it. What he gets in return is even greater – seeing Kiefer's paintings in real life. They are enormous, many meters high and wide. The size makes them very vivid.

I did not know about Anselm Kiefer at all, but I am completely enthralled by the paintings. I cannot explain how other than they touched me deeply. Especially two of them. The largest is called *The Land of the Two Rivers*, a landscape from our origins in blue and yellow colours. The second, *The Renowned Orders of the Night*, shows a man lying on the ground looking up at the stars. I recognise myself in that man – like when I sit on the yacht and look at the night sky filled with stars – in wonder.

We hadn't planned to visit Getxo. We ended up there because it was a convenient distance from Santander. I experienced it as another nice gift from existence. I got a role model for our upcoming trip – that resting man on the ground.

What is your experience with giving and receiving?
What is the easiest for you?

18

THE VIRGIN TOUR

"You can never cross the ocean unless you have the courage to lose sight of the shore." – Christopher Columbus

We are in Cascais, a small and beautiful village outside Lisbon, in Portugal. *Vista* is filled to the brim with food, some should be enough for New Zealand. From now on, we only have small islands in front of us. The first will be, Porto Santo – a small island north of Madeira, 500 NM southwest of us. For the record, for Arthur, this was no big deal as he counted the 350 NM from Falmouth to La Rochelle, in Biscay. For me, it was still close to the coast.

It is exciting and terrifying to leave land out of sight. The first time we did it was when we sailed to Gotland, 50 nautical miles from Västervik to Visby. We waited for days before we dared to go. This time we have a much longer journey ahead of us, and we are waiting this time too.

The special thing about oceans is their enormous size. The waves from far away are called swells. Then there are waves, usually smaller, created by the surrounding wind. The two types of waves can come from different directions, I will learn.

We are lucky to get a good first passage. Three days later, we drop anchor in turquoise water and sand at Porto Santo.

A Historical Place

We soon learn that there is a party going on. They will celebrate the return of Christopher Columbus. Still. I have to read up. In 1492, he set out for the first time to reach India by sea – something he had wanted and planned for over ten years. At that time, it was known that the earth was round, but not everyone believed it. And you would see his charts, like a sketch compared to today's charts. No wonder they were scared and that it was dangerous.

His flagship *Santa Maria* ran aground in what he called the West Indies. That did not stop him from continuing, he arranged financing for new ships and voyages. A replica of the *Santa Maria* has been built with a home port in Funchal on Madeira. Every year, she gets to take a trip to Porto Santo, where Columbus also lived for a time.

To my surprise, I am very touched by the ceremony. The *Santa Maria* sets out with full sail from the port of Porto Santo and drops anchor on the long sandy beach outside the village. Captain Columbus and his closest men are picked up in a rowboat that takes them ashore. On the beach, dancing and singing women meet them and offer food and wine.

We felt the wings of history. Soon it will be our turn!

Almost to the day, five years later, we drop anchor again at Porto Santo. Our circumnavigation is complete. It is something to celebrate, I now feel!

We are going on an adventure! Sailing the sea is an adventure for everyone, whether it was a hundred years ago or now. It is exciting!

What would be an adventure in your life?

19

ACCEPTANCE – BEING A NEWBIE

"It does not matter how slowly you go as long as you do not stop." – Confucius

The theme of learning was particularly present at the beginning of our journey. Below is an article I wrote six months into the trip after I had completed my first outdoor dive in Madeira.

The Power of Learning

My last year has been full of changes. The biggest thing is that my husband and I have started sailing around the world. We have sold everything and live aboard our sailboat. We have chosen a completely new way of living. It feels unusual, we have a thousand questions about everything and learn new things all the time. I have been given many reasons to think about how central and sometimes uncomfortable lifelong learning is. It is important to surround myself with people who believe in my abilities.

It is nice to feel competent – I left a life where almost everything was perfect. I appreciated my role as a master certified coach and self-employed person. There was a lot to take care of, and I enjoyed it. I felt that I knew my profession well after twenty years. Even though it was new to run a business and find new clients of my own, I learned that too. I had found an office in town that I loved, and my clients were making great progress.

What if I had let myself stay there? I who love Stockholm. It had been so easy – everything worked so well. When it was the most comfortable two summers ago, life took the opportunity to knock me over and make me wake up and set up a new, bigger game. Now here I am – 57 years old and feeling like a kid again – in a sailboat on my way to the other side of the world.

The hardest thing to learn so far has been the diving. It is not something I dreamed of learning. We both needed a diving certificate to clean the bottom and cut fishing nets – for pure safety reasons. The diving school gave nine completely different, more lustful, reasons.

Right from the start, I had big problems emptying the mask. I had to start by watching as the others trained further. It made me so disappointed and angry – we had so much to do, and I did not have time to redo it. But it was just to accept the fact and take new steps. I did pass at the very last minute. Thanks for the great help from my brother and a very patient dive master. We did our mandatory open-air dive in six-degree water and a dry suit.

Now, six months later, I have my first real outdoor dive in warm water behind me – immortalised in a short film! I have got a glimpse of a whole new world to discover.

On the way to getting my PADI certificate, I got a lot of encouragement. Not one person commented or had an opinion that it should be faster, easier or different in any way. Except myself. I, who coaches others, should... No, in this case, I thought about how fumbling and insecure I must appear to others, so I kept silent about my slow progress. I wanted to look good and be competent and confident all at once. So devastating and unnecessary.

I began to think about how courageous it is to dare to take on new things in life. All training and willingness to learn new things was a prerequisite for us to dare to jump into a completely new lifestyle without a financial parachute. I understood that this willingness was also the reason why we were able to move from decision to action so quickly.

I look at the groping steps with new eyes. How brave it is to tremble and be vulnerable in front of others. I see the determination and will it takes to decide to learn something new, to go from words to action. Like a Bambi on slippery ice, who does not care if others will laugh, understand or help.

From now on, I applaud all new learning — especially the difficult and painful first steps. It is at the beginning when self-confidence is at its lowest and uncertainty, self-doubt and the risk of failure are at its greatest – that is when we need help the most. I have to put up with the fact that not everyone understands the big things that are going on inside me. And then one day it happens – then I have cracked the code, found the balance, got over the barrier. The worry is as if blown away, I hardly remember it.

I send gratitude to all my wise friends who have already understood this and who have been an important support for me.

Imagine the difference it would make if we all honoured the person who takes on something new to learn, regardless of age. What if we remember that we have managed to learn new things many times in life – a strength – something we can do again and again. What if we were all a little kinder and more patient and accepted our own and others' shaky steps and remembered that this is exactly how we grow. Yes, what would be possible then?

All the best from Anna

La Palma, October 29, 2019

I can talk for a long time about the importance of learning new things throughout life. For me, for us, it has become a way of life. I am surprised that awareness is not higher about how important it is to be open and curious about new things. It is a must when it comes to development.

How do you treat yourself when you learn something new? And others?

20

THE SECRETS OF THE NIGHT

"It is during our darkest moments that we must focus to see the light." – Aristotle

Before *Vista*, we had never sailed a whole night. I knew, of course, that there were lighthouses to show the way. Beyond that, my knowledge was as black as the nights could get. Arthur made a comparison with land and assured me that the road is there even when it is dark. The only difference is that we do not see that far.

With the night sailing, a new world opened up for me. I discovered that there was a lot out there that showed the way. In a way, it was even easier. The little we saw was exactly what we needed to see to sail or safely enter a port.

The Moon

In Stockholm, we rarely saw the moon, only other houses. Out at sea, the moon becomes very visible, like a big light bulb, giving a lot of light at night. As a child, I had a night light, a light bulb that looked like a mini-moon – very soothing to look into. So, it is out at sea, too. It is like having a friend, a companion, following me through the dark.

We quickly learned to check which moon phase we were in before a longer passage. The best thing was if we could time it, is to sail with a rising moon. It feels like we are distinguished guests of the sea when we sail on the silver carpet of the moon.

I did a photo odyssey on my lunar year, from Stockholm to New Zealand, including my first sight of a red moon. It is published on Medium.com and Noonsite.com.

The Stars

The Vikings navigated by the stars – all the way across the Atlantic, already in the 11th century. The pole star shows the way north. Venus shines brightly in the morning and evening. We sail west with the Three Wise Men/Orion's Belt in sight, night after night.

In the past, a sextant was used to find the position using the stars. Nowadays, we get this information via satellites, and apps, like SkyView®, which shows us the names and positions of the stars and planets.

The stars point out the larger context we are part of and reminds me of the opportunity to see further and bigger. The stars are numerous and illustrate abundance. They are grouped into constellations, like a family or a company. My inner guiding stars are my values.

Navigation Lights

At sunset, all yachts and cargo boats must light their navigation lights, even if there are no others nearby. I light ours and usually think it is my way of saying hello to the stars. The different lights give us information about the type of boat we are meeting – whether it is sailing, powered or anchored. The green light shows the starboard (right) side of the boat, and the red light shows the port (left) side.

Fishing boats, unfortunately rarely use AIS, so those are the ones we watch especially closely. They use a bright white light when harvesting their nets.

AIS, Chart Plotter and Radar

AIS is an automatic identification system that shows on our plotter where boats are in real time. For ships in category A, commercial traffic, it is mandatory. For recreational boats and fishing boats, class B, it is optional. I remember when it came – some did not want to show their secret anchorages, or where they were during their holidays. Out on the seas, there is no doubt – we want to be seen as much as possible. During darkness, AIS becomes extra important. We can see a ship on the plotter around twenty nautical miles ahead, compared to the navigation lights, which are only visible a few nautical miles.

The other useful instrument is the radar. It shows rain, waves and ships on a screen. On the pattern, we can see if something unfamiliar is in our way, or how big or in which direction something is moving.

Lighthouses

When we get closer to a coast, a lighthouse is often the first thing we see. A fixed or flashing light – unique to each lighthouse, shows the way. The white sector shows the safe path. Entering the red or green sector means we must adjust the course.

Leading Lights

Leading lights have a similar function to the lighthouse. Here there is a bright white light from two positions on land – where they meet, they form an even line, the right sector to enter a port. At night, it is excellent guidance, as many other competing lights are mostly off.

Light and Sound Buoys

In high-traffic areas, the buoys have either lights or sounds and sometimes they are also numbered. The first time I heard a sound buoy was off Helgoland, a small island in the North Sea. We got there at night and were quite nervous to find the small marina. Once we understood that they were using sound buoys, it became so easy. One by one they announced themselves. We slowly made our way towards the marina, where we could dock as the fifth yacht out from the pontoon.

The Magic Things

At nighttime, it is also possible to see magic phenomena like "milky seas", or Mareel – a luminescent glow in the surface of the water. Birds and dolphins also come at night.

There is so much to explore – in the outer as well as the inner world. I like the word *world* – we all have our own world – secret to ourselves until we start exploring it, and secret to others until we start sharing ourselves.

What world are you curious about now?

21

FEAR

"At nadir – the darkest point at the proverbial, and often literal, bottom – the person is also closest to the divine." – David Drake

It is pitch-black outside. I just got up for my night shift. No moon and no stars, a cloudy night. It is windy; I hear the wind whistling in the rigging. It is at night that fear can start to spin – what is out there? A whale? Fishing nets? A drifting container? Impossible to see with my eyes on a dark night. A yacht without AIS, and not lit navigation lights (it happens, they might want to save electricity, or the lights might have gone out) I would see up close, but even that may be too late not to collide with each other.

Fear can pop up all the time. Fear is about the unknown future, for example, the next part of the journey. For some reason, the sea in front of me always seems to be the trickiest. What could happen in the worst case?

Embarking on a new journey will contain fear and new challenges. We will be tested if we are ready and have the necessary courage. David Drake, PhD, uses rites of passage as a metaphor for this kind of transformative journey. Moving from one stage to another means seeing our shadows and facing our fears.

The Guardians
"The guardians can often be seen as projections of the fears and aspect of our Shadows that are critical for the crossing of this threshold. Terms such as

"good" and "bad" are not relevant here because every guardian serves the same purpose: to lift the veil that has kept the person from seeing what was there all along and help him cross the threshold of truth. Every guardian essentially asks the same questions:

What do you need to confront to gain access to what you are seeking?

What projections do you need to withdraw so you can bring that energy into yourself to support your growth?

Guardians help to see more clearly, sink deeper into truth, provide aid, test readiness, and protect us and the threshold. They keep the boundaries in good repair and the center fresh and strong.

Guardians are stewardesses of vital crossings and of the deep grief and deep joy that often are found there. They know that when some part of us dies, there must be a time of mourning, a period of withdrawal and introspection, a period of allowing tears to fall."

(David Drake, *The Narrative Coaching*, 2015)

To Face the Fear

For me, fear shows me the way forward. I know when I face my fears, I will expand and get in touch with my resourcefulness. The fear acts as my inner barometer.

Living and sailing with both great risk and fear makes me alert and present to the moment. Anything can happen. I have to realise that I cannot control, or know, how the future will unfold. I have to trust that we also belong and are taken care of. The sea and the waves carry us, no matter how high the waves get. Often the toughest situations are the most memorable. Afterwards. They show me that I/we can do this too.

It is important to take care of the body – not to fall, to get enough sleep and to have enough energy. Being aware of the body is a great way to become present. Following each breath, for example, or picking up on what my senses are telling me.

I know that I let fear rule when my life is stagnant and routine. That is why I choose the new as often as possible. When I am afraid, I do what I can to be present. Everything passes.

Fear
"It is said that before entering the sea
a river trembles with fear.
She looks back at the path she has travelled,
from the peaks of the mountains,
the long winding road crossing the forests and villages.
And in front of her,
she sees an ocean so vast,
that to enter
there seems nothing more than disappear forever.
But there is no other way.
The river can never go back.
Nobody can go back.
To go back is impossible in existence.
The river needs to take the risk
of entering the ocean
because only then will fear disappear,
because that's where the river will know
it's not about disappearing into the ocean,
but of becoming the ocean."
– Khalil Gibran

What are you afraid of?

22

THE BIG JUMP

"Nothing is impossible. The word itself says 'I'm possible!" – Audrey Hepburn

A sailor, Nisse, who has crossed the Atlantic before, visited us while we waited for a good departure day in Las Palmas, Gran Canaria. He teased us and called it "a jump across the puddle". That is not how I felt about it. I was most worried about the food; could I cook on the open sea? And again, for the waves – how high would they be?

In addition to these concerns, my boat neighbour Anette, and I engaged each other in the risk of robbery in the Caribbean, something we had heard was common. We discussed many options on how to defend ourselves. Neither of us could see ourselves making a physical defence, so we laughed with relief when we finally came up with the peaceful plan of handing over money in a child's fishing net and saying, "Here you go, please, leave us now."

Many years before, maybe at nine years old, I was standing trembling on the edge of the five-meter diving board when it was my swimming teacher Harry who shouted "Jump!" And I did, in pure trust in him. Now it was high time to jump again. And one day we did – jumped over that puddle.

Below is how I described our crossing immediately afterwards.

Crossing the Atlantic – Yes, We Made It!

Since Columbus wanted to sail west to India, and instead found land in between, many have sailed to the Caribbean. We are just another yacht sailing with the trade winds. No matter when – the first time is always the first, and it is not done until you have sailed across.

So how was it for you?

We loved it, we felt like we lived out there. We could have sailed on for many more days.

It is magical to be in the wind and waves for so long. We were surfing a massive wind and wave movement. We did not see a boat the whole trip, only four on the chart plotter. One of them a rowboat! The context is huge – horizons all around and many thousands of meters of deep dark blue water below us. At first, it was a bit scary – so much can happen along the way, and there is no going back once you set off. Pretty soon we felt it was working – the waves carried us. And when the wind picked up, we just furled our Genoa in a bit. The beauty and strength of nature took over.

We are still in awe by the fact that we have sailed, the two of us across the Atlantic. We arrived in Martinique at dawn, and it felt like entering a new land. The light was different, and the turquoise water was warmer (28°). The view was also different – a huge bay with many sailboats at anchor, surrounded by soft green hills. Absolutely fantastic!

It is special to travel so long to reach a place – it takes less than a day to fly here. For us, it took two weeks from Cape Verde, or six and a half months from Stockholm.

Did everything go well?

Yes, we have a wonderful yacht, an Amel Super Maramu, built to sail across the oceans. She sailed very nicely in the deep waves. We are going to have someone look over our Genoa (the only sail we used) as parts of the sunshade started coming off at the bottom of it. We also had some chafing on our sheet – a beginner mistake – we have bought new ones in reserve.

How did you manage, just the two of you?

Since we have a yacht that runs on autopilot, it is more about staying fit and rested for the night watches. The sailing itself was fairly easy, with the trade wind blowing from the northeast. And as mentioned, no boats out there, so not much to check out.

But you never know! We checked the radar every half hour and on the latter part of the trip we got squalls – strong wind and rain for a short while, mainly at night.

Constantly rolling was the hardest thing to get used to and made us very present to every single movement. It also had us adjust and secure things that moved around and sounded. The evening before departure, we both applied seasickness patches. They last three days, which was enough for the whole trip.

We are very aware of the importance of sleep and rest, so we prioritise that all the time. We have a fixed schedule for the night watches. Three hours, starting 22-01, 1-4, 4-7, 7-10 in the morning. If we cannot sleep, we rest. We always take at least one nap a day. Just in case, we have sleeping pills too (never used).

I cooked a few days in advance, so that neither of us would have to cook until we got used to the waves. We have a pressure cooker with a lid that cannot come off. That is perfect, as everything in the galley must be guarded. We also brought dry food (like for campers), added a little warm water to and let stand for ten minutes. Also, that as security. We did not drink alcohol, just lots of water, coffee and tea.

What do you do all day? Is it not boring?

The days go by fast! Not boring at all. We live on board. We want it to be as smooth as possible every day. We are in no hurry, and prefer slower speed, if it increases safety.

Of course, we check the weather, and everything related to the yacht many times a day. We adjust the sail if necessary and twice a day we make fresh water, hot water and charge the batteries.

Arthur and I approach it differently. I have various projects going on – writing a diary and other texts, reading, taking photos, listening to podcasts. I choose the thickest and deepest book I can, as I have presence and time to read quite a lot.

On this trip, I studied the sun and the waves every morning and evening. I sit in the front row and drown in the big silent light show. I am with everything that is around us. It is kind of like being on a quiet retreat in a magical setting. At night, the moon shone above us. I am learning to trust, especially the big waves – they carry us with care even as they grow taller and stronger. I witness and see more and more. I am very grateful.

Facts

Start December 19, 2019, from Las Palmas.

A stop in Mindelo, Cape Verde, for a few days.

Started again January 1, 2020

Arrived in Martinique January 15

Distance covered: 887+2108 NM (total 2995 and from Stockholm 6517 NM)

Engine hours: 25

Wind speed: 15-30 knots, mostly around 20

Wave height: 2-4 meters

Speed: 6-8 knot

I think it is worth reminding myself that it is a matter of mindset whether something is big, scary, and difficult or not. Now that I am back in the same place from where we started, I feel much calmer, but also humbler. What is super clear is that worrying does not help, it just does not.

What is your current box?
What would be a big jump for you?
What are your thoughts on your capacity to make it happen?

23

IN SHOCK

"Feelings are just visitors. Let them come and go." – Mooji

In a state of bliss, we remained at anchor in St Anne Bay, Martinique, for many days. We enjoyed relaxing, admiring the turquoise water and the sun going up and down over the low hills. We did not even launch the dinghy.

One afternoon the wind picked up, and I adjusted the fenders – and lost one. I watched it disappear out to sea. Arthur was in the cockpit and jumped in to save it. He managed to get hold of it, but then the tough part began – swimming back against the strong current.

I stood on the stern and looked at him. In paralysis. I did not throw out the safety line. I did not turn on the VHF and call for help. I did not even wave my arms. No one would see it anyway, since we were far out with no other boats nearby. I just stared at Arthur. As if that would help. He came closer. It took time, a long time. When he finally got close enough, he yelled at me to throw in a line. I did that. Luckily, we had the bathing stairs down and Arthur managed to climb up on his own. He shivered and sobbed, cold, scared and relieved at the same time.

This was a huge lesson for us. I felt so guilty for not doing anything. In shock, I understood afterwards. We have saved fenders after this, but not by jumping into the water. We decided that staying on the yacht is always a rule. We cannot count on rescuing each other.

The guilt was so strong, it took me a long time to even write this down. I remind myself of the important question: how did you notice he was okay?

Have you ever been close to death? If so, when did you realise that you, or the other person, would survive?

24

THE PULSE OF THE PLANET

"Not everything that can be counted counts, and not everything that counts can be counted." – Albert Einstein

This is a text I wrote on our first passage across the North Atlantic.

We fall into a new pulse – slower and larger, more erratic than the seconds, minutes and hours of the clock. The pulse of the planet manifests itself in the waves, currents, high and low tides, the winds, the sun, the stars and the moon. A rhythm that is infinitely greater than what I as an individual can control and imagine.

Imagine the waves – here in the Atlantic – they are bigger and longer than in the Baltic Sea. They crawl in a common mass up to a wave crest, to jump off and relentlessly continue the next climb upwards. With joint power, they move forward.

I see the crests of the waves build up in the stern and feel the *Vista* follow up and down. I remembered the first time we sailed on surf waves, from Bornholm to Simrishamn. I only glanced back once at the big waves. After that, I took a firm grip on the steering wheel and shifted my focus on what was ahead. Now the waves are even bigger, and I practice trust – that the waves will carry us. My task is to follow along, to be awake and present to the pulse. Having the courage to expand and open my mind even more.

My eyes seek the light. The sun breaks through the clouds and sends down its rays in a large curtain. I never get tired of the scenario: sunrise, sunset, sun glitter and sun streets. The body feels it too – several degrees warmer immediately.

At night, I wait for the moon. She shows herself far from always. It is because it is cloudy. But the clouds are not always visible at night either, so I wait and hope. The moon is very erratic. New Moon or Full Moon – that cycle feels tangible, but otherwise… She can appear anywhere in the sky, in any direction, be big or small, white or gold-coloured and appear for a tiny, tiny moment or, if I am lucky, all night. Regardless, I am very intrigued by this night light.

The stars are still like sprinkles in the sky. Eventually, we will start using our sextant, to find out our position with the help of the stars.

We have met a sailor, Max, who long ago, at the age of 21, navigated with the help of a sextant on his journey around the world. He said that sailing across the Pacific would change us forever. That makes me even more curious. What will happen? How is it possible to describe and perhaps even justify a journey of this kind? So far, it is still quite inscrutable. I have thought several times that I could die now. Get me right, I want to live. It is so big and powerful, everything. I feel extremely grateful and hope to share it with these lines.

The rhythm of nature is strong and obvious when we are right into it, with nothing else distracting. For me, it was a good exercise in accepting and taking in all the energy that flows and pulsates within me and around me.

Can you relate to the pulse of the planet?

25

BUOYANCY

"The more you are aware of your buoyancy, and therefore your breathing, the more everything will flow peacefully." – Elisabeth Hoeffnagel

Bonaire is called the Divers' Paradise. The small island has eighty-five dive sites and has worked deliberately for a long time to keep its corals and underwater world clean and alive. No boats are thus allowed to anchor, instead they have buoys. We stayed for a month and took the chance to practice some more diving. We were lucky to find the dive master Elisabeth, who understood our situation – that we must be able to dive on our own, at least for safety reasons. We did the PADI Buoyancy Course with her, which was perfect. With my rough start to diving, I was very happy to manage the buoyancy on my own. This is how I wrote about it afterwards.

Buoyancy – When You Have It, Everything Is Easier!

Do you remember when you first started learning to ride a bike – the feeling of finding your balance? The wobbly start that ended in falling, and then suddenly did not. You realised how to hold your body and pedal around and around, to get a more and more stable sense of balance. You got it!

Same thing with buoyancy for a diver – without it, you cannot dive, it is too risky. I have eleven kilos of air on my back and seven kilos of weights on my hips to compensate for the air. I am about to find the buoyancy to dive safely and effortlessly. It gives me access to a positive interaction with the underwater world. A world with its very own premises.

You have buoyancy when you can float in the middle of the sea without holding on to anything. Especially not the corals. You fine-tune by breathing – in, and you go up, and out, you go down. "If you have buoyancy, the rest is easy", as our diving instructor Elisabeth said.

So, I struggled. Floated up and down, gasped for air, flapped my arms, paddled my feet/fins, touched the bottom – thankfully mostly on sand (and yes, I touched corals, I am sorry, please forgive me), was too fast and clueless about how to manage myself.

Elisabeth checked all external things like BCD, the correct number of weights, where I had the weights and so on. She was like a scientist looking for problems and then coming up with solutions. The last thing we did was breathing together.

Then came the moment when I felt buoyancy! I felt joy for many minutes, before I floated up (again). Anyway, now I knew what I was looking for. I have cracked the code for how to get back to it with the breathing and make the movements smaller, more streamlined. Now, it was time to glide through the water like a fish.

*

Arthur and I were talking the other night about how important it is to have buoyancy, even on land, in our communication. When we are in balance, we are aware of our breathing, and we see our surroundings. We are relaxed and present. We can easily get back to buoyancy by sharing what is there.

For example – I meet a new person and immediately forget his or her name. My breathing is shallow, I am nervous. By saying what is there, instead of trying to maintain a façade, I give myself space to be authentic and to reconnect. Both with myself and the person in front of me.

The same when we talk – we are present and aware of our own and the other's state – it gives a nice flow to our conversation. We listen deeply, have space for the other person's thinking, feeling and talking. No need to steal the air and interrupt, because we are in no rush. We rest in the listening and know that much is happening, even if it is subtle. We hear the nuances. We dare to wait and be silent. Waiting curiously for what comes next. You never know what is coming.

We trust that we will be listened to as well. We do not have to think about our answer or our story in advance, we just listen. When it is our turn, we rest with ourselves too – giving ourselves space to just be and allow thoughts, new ideas and answers to emerge.

Wise Elisabeth again, "The more you are aware of your buoyancy, and therefore your breathing, the more everything will flow peacefully. In the water and out the water. Stress because of new and unexpected things happening, will vanish, and your worries will disappear."

As in emergencies on the plane – secure your breathing first. As we breathe, we will feel more – what we want, need and what is most important at this moment. How aware are you of your breathing? I was not, I must admit. I thought I was breathing deeply, but I did not, at least not underwater – a new environment for me.

Breathe in, breathe out – easy!

Enjoy your day!

Sun and salt from Anna

Bonaire, 23 February 2020

Being buoyant is subtle and requires a high level of awareness of both the inner and outer states. Other ways of expressing the same phenomenon are being in sync or having a rapport with each other and the environment. As always, we have to start with ourselves. A lifelong subject to practice!

Are you good at listening to your inner state?

26

WE GOT TESTED

"A problem is a chance for you to do your best." – Duke Ellington

It is exciting to arrive at Shelter Bay Marina, our final stop before the Panama Canal, after months of preparation. Since we lost the bow thruster propeller when we left Curaçao, we have been given special permission to dock at an outside pontoon.

We are also grateful to be here, as it was very close that our journey ended in Curaçao. A large motorboat full of partying people lost control and crashed into several boats in the small marina. Our prize was the davit, which prevented a bigger hole in the yacht. The harbourmaster prayed for us and managed the negotiations, so well that it resulted in more compensation than the insurance company would cover.

*

As always, we start by checking in. Our agent, Erick Gálvez, meets with us and sorts out all the paperwork with the Spanish-speaking authorities. He is the one who organises the payment and that we get extra fenders, long lines and four line handlers. But first, a happy guy measures and checks *Vista* and takes a selfie with us as a finale.

Arthur now has to do what a man has to do – dive and screw on a new propeller (we had one in reserve, thanks to the captain). The only catch is that two crocodiles are living in this marina. The harbour captain, Juan, told us that he would never go in the water again after saving his skateboard. When he dove to pick it up, he got into a fight with the crocodile!

With seriousness, Arthur puts on the diving gear and prepares to descend. We notice how a "marinero" (marina crew) quietly positions himself further out on the pontoon without us having asked him to. I am on the pontoon to give Arthur one screw at a time – six or eight in total. The relief is great when Arthur is ready after thirty minutes.

A Big Decision

Only now is it starting to become real how big and irreversible the decision to go via the channel is. We have had New Zealand in mind, but now I realise this step also includes all the way back to Europe. Too big to embrace, until now that we are on site.

We have experience in locks on canals. In Sweden, we could get from the east coast to the west coast via two canals in a week. The Göta Canal consists of fifty-eight locks that start at Mem, Söderköping and end at Sjötorp in lake Vänern. Then the Trollhätte Canal takes over with six locks that take us on to Gothenburg – or vice versa. And we have gone through the Kiel Canal – a big deal at that time.

But this is different. You cannot get through the Panama Canal on your own, even if it only takes a day or two. Every pleasure boat must have an official co-pilot through the canal. Another rule is to have four people, besides the captain, responsible for handling the long lines to be thrown up onto the high quays. I did not want to be one of them, plus seven people would have real cooked food, not just sandwiches, according to the strict guidelines for everything related to the transit.

And Another

After five days, on Friday, March 13, we got our slot time for the channel. The lines and fenders were already on board, and the line handlers were due to arrive later in the day. That same morning, we received a message from Erick to come and see him. A little worried, we went up to the marina restaurant where we saw him busy talking. He quickly broke the news: A lockdown will go into effect starting tomorrow. Do we want to go to French Polynesia, or not?

It was a big decision to make right away. A catamaran with a family on board, with whom we were supposed to be nested with in the channel, decided not to go. An Amel sailor chose to sail to Hawaii, instead of French Polynesia. They had heard that French Polynesia was closing.

We decided to go. We followed our motto, to go wave by wave. Because everything for the next step was ready. The most important thing was that we had the "zarpe" – the approval document, to leave Panama for Nuku Hiva in French Polynesia. Our slot was still open, and our agent was willing to check us out. So, he did, and we left Panama after filling the boat with food from the excellent open-air Panama Mercado. It would be perfect to be in the Pacific during the lockdown, we thought when we predicted that it would be over after a month. The next day they closed the canal.

The lockdown was a key moment where our commitment to sailing to New Zealand got tested. Another decision from us would have changed the whole story. Now we know it was not over in a month. For many, these years with COVID-19 became very challenging. We trusted that we were meant to do this trip, saying no was not an option.

What is your memory of following your truth and commitment, despite low chances of success?

27

THE BIG BLUE PACIFIC

"A man is never lost at sea." – Ernest Hemingway

The Pacific, covering almost half the world, was a great mystery to me. In Stockholm, I looked at Google Earth and saw all the blue covering the other half side of the world.

Not even New Zealand, our primary target, was visible. "Zoom in", Arthur said. And I did, more and more and more until some tiny little islands appeared in the middle of all that blue. Arthur knew, he had read the cruising guides for many years and followed other sailors, like René and Paulien. He had already travelled to French Polynesia many times in his dreams.

For me, the big blue represents an open, empty and silent space. Even though we had a long passage behind us, this one was longer and meant other important opportunities. I thought of Otto Scharmer's Theory U, where one must first leave the past and established behind. In the void, new opportunities come to connect with the source and create something new. My grandfather Elis, who is an artist, called the same space the Immanent Sea – a fitting name for us. And I also carried with me Max's words – that the Pacific passage would change me forever. It felt devoutly big and important to experience the void. What would happen out there in the great infinity?

After a few days, I started writing a daily reflection and chose an image that represented the day. I came in contact with the light of our planet. I loved being out there. You are welcome to join me on my journey across the big blue. You can read my Pacific Reflections on Medium and Noonsite.

Shocking News

Every morning and evening, we download the weather via our satellite phone. It takes time, so it is a bit of a procedure. We can also receive short satellite messages. Somewhere in the middle of the trip, we got word that French Polynesia was closed. The government required us to sail directly to Papeete (one more week from Nuku Hiva), leave the yacht and fly home. This was a shocking announcement, amidst all the beauty of sailing the Pacific. We sat in our only home and looked at each other, not knowing what to do.

We have 150 minutes for emergency calls every month on the satellite. I called my brother and asked him to check if there were any countries, in this big blue, that were open. He reported back: Hawaii and American Samoa, around 4000 NM to each. Arthur and I looked at each other again and considered the option of turning 90° north instead, as our sister yacht had previously decided. We were not afraid of the distance, a month – we were into the ocean sailing now and loving it. But it did not feel good, so we continued. I informed the government in Papeete that this is our home, and we cannot fly home. For that, they replied, you must go elsewhere.

Luckily, I misunderstood and informed the JRCC at Hiva Oa that we were coming to them instead of Nuku Hiva. They welcomed us and I informed the government of our new destination. He explained that he had meant somewhere else, outside French Polynesia. If you look at a map, you can see how long all the distances are, and as far as we knew, everything except Hawaii and Samoa was already closed. Maybe he realised how difficult it was because in the end he said yes and let us go to Hiva Oa.

The big blue can represent everything that is unknown to us. Another metaphor is the iceberg, most of which is hidden. Everything new is unknown first.

How do you feel about being with the unknown?

28

INSECURITY

"The man of understanding will accept that INSECURITY is the very fabric of life, and that NOT KNOWING is the counterpart of the miraculous and the mysterious existence." – Osho

Various topics came up in my daily reflections on the passage of the Pacific. After three weeks, I wrote about insecurity, a theme that has recurred many times.

Friday April 10, 2020, 07°30'S, 130°22'W
Marquesas Fracture Zone

It is Good Friday and Easter – the day Jesus was crucified over 2,000 years ago for his, at the time, provocative speeches. What an impact he made. His speeches still serve as the foundation of a major world religion.

Who comes next?
Who will bring a new vision for the world?
Who has answers to today's big questions?

Looking out at the calm ocean.
Wind and waves have gone down.
We are slowly gliding forward.
Sun is making a broad road of glitter for my eyes to rest on.

I am thinking about how nature solves things
– cyclones for example.
There is a Harold running around, south-southwest of here. People living
here must rise again and again
like we all did when we learned to walk.

The whole world is currently both united and separated
at the same time, due to the deadly virus COVID-19.
Most borders are closed.
No one knows when and how this will heal.
We learn to keep our distance,
to not hug, kiss or shake hands.

Who will remind us to trust each other again?
Sun says: You will.

☼

This state of emergency for the whole world reached us even in the middle of nowhere. The interesting thing was that existence, and its beauty were much stronger than the fearful messages we received via Iridium. The uncertainty I felt from time to time was swept away by the waves and the light. For me, it was proof of how magnificent existence is.

Who reminds you to have trust?

29

TO BE RECEIVED

"And, when you want something, all the universe conspires in helping you to achieve it." – Paulo Coelho, *The Alchemist*

After thirty-one days in the Pacific, we saw land, the high mountains of the small island of Hiva Oa – one of the Marquesas Islands in the easternmost part of French Polynesia. The same day we anchored, I wrote the text below.

Please Receive Us Well

Never before have I felt the true meaning of the yellow flag. When you enter a new country, you must declare that you are free of infection. You do this by raising the yellow quarantine flag. Until today, the flag has only been a formality, a way to show that we are new and want in.

With COVID-19, it is a whole new story. Already in contact with JRCC, forty-eight hours before going in, we needed to confirm our health status. We know that on these remote islands, there have been people from the West who have brought diseases in the past. We know they do not have the resources to heal anyone who has been affected. At Hiva Oa, everyone had spent thirty days in quarantine. I understand that they are afraid. I will do my best to make friends.

Go Home

They were not afraid, we soon realised. We were the scared ones. The words "Go home" and "Go somewhere else" still rang in our ears. Even though we had got permission to enter, we had no idea what that would mean in practice.

Trust Me

So, we were shaky about what would happen upon our arrival. Only Arthur got allowed to go ashore. He met Mark, the head of the JRCC, a teacher and now the contact person for all the sailors. There was also a young woman, Hereite, who translated French into English, and two policemen. Arthur told our story and even cried. He was afraid they would send us away as soon as we had done the bare minimum to continue to Papeete – and fly home.

Mark calmed him down and said, "Trust me, you can stay here. I'll take care of this. Trust me." And he did – Mark wrote a letter to the government in Papeete describing how old and tired Arthur was, that we had to stay. It worked. After a few days, Mark called on us both. This time with a new message, "You must stay here; you are not allowed to leave." With a big smile on his face, he asked Arthur how it sounded. Our gratitude was immense.

We later understood that the local people have their way of dealing with things. They do not like the French government controlling them in detail. For them, it was very natural to receive everyone who came to them after always a long journey. A completely different message than from the government.

The Gift

After another week of quarantine, all sailors, about thirty yachts, were welcomed ashore for the first time. The locals had made large boxes of fruit for us, one per yacht. Filled with local exotic fruits like: Pamplemousse/pomelo, papaya, passion fruit, avocado, bananas and star fruit. The governor of Hiva Oa, a lady, spoke to us and thanked us for being so loyal to the restrictions. For us, it was a very short time, for others it meant over forty days.

Then it opened, step by step. First, we got to sail around the Marquesas

Islands. They had made a special flag for us all, showing that we had permission. And then the whole of French Polynesia opened up.

We were very happy and grateful. It was again so close to the end of the greatest journey of our life. Thanks to the people at Hiva Oa, that did not happen.

Making a landfall is always big. Even in familiar waters. In unfamiliar waters, it is exciting. Being welcomed is fundamental. Hiva Oa will always have a place in my heart. Here I got the feeling of coming home.

Where do you feel at home?

30

TO EMBRACE CHANGES

"When you change the way you look at things, the things you look at change."
– Abraham/Esther Hicks

In my profession, I often met clients who were worried about change. Organisational changes in large companies, come so often that someone came up with the acronym BOCHA: Bend Over, Here it Comes Again. In our VUCA world: Volatile, Uncertain, Complex and Ambiguous, we all need to develop change skills. I, too, had models and ideas about how we could strengthen our ability to face change. My understanding of change has deepened a lot.

To Be Humble

Man may think that we are the ones who control this planet. Being out sailing the oceans has made me humble before Existence. Nature and the weather surround us all the time. The walls of a boat are never thick enough to exclude forces and movements from nature. We feel and hear them all the time, 24/7.

As a sailor, it is a must to embrace change. I have to accept and follow reality, the forces of nature. It is super clear that nature, existence, is constantly moving. It is a sign of life to move.

Weather

An important part of being a sailor is to dance with the weather. If we look at the whole world from above, we will see all kinds of weather going on at the same time. People have figured out that it is possible to sail with the wind and the current around the world, at certain times of the year. That is the play we are after, sailing down-wind around the globe.

We wait for the best conditions, for weather windows. If that means we have to wait, we wait. We also know that it is colder closer to the poles and warmer around the equator. We are drawn to sailing in warm blue water. When we are out there, we are alert and connect with the wind, and the waves, as best we can. We adjust the sails, to show that we notice what is going on around us.

It Always Changes

Of course, it is not always blue sunshine. But the sun always comes after rain. The wind stops blowing hard after a few hours or days. Rain never lasts forever, nor does fog. When the cloud is empty, it will dissolve. The sun always rises and sets (as we see it) and so does the moon. Watching and being in those changes every day, makes me realise that everything is always changing. That is life.

There is a saying that a sailor's plans are written in sand. When morning comes, the tide has washed away the words, and new plans must be made. Zen meditators consciously make paintings in sand, practising acceptance of the ephemeral and perishable.

Saying Yes

Changes do not only come from existence; I also want change and variety. Whether the changes are self-chosen or not – the question is how well do I handle them?

As a sailor, I have learned to accept change and to say yes, quickly. I practice seeing the changes as gifts in the moment. It is not always easy. I like to plan my days, and suddenly having a few hours or days of free time can bother me before I see the opportunity in it.

It's Over, It's Over!

Many years ago, we were in Gedser, a small village on the southernmost tip of Denmark. It is a strategic place when one wants to continue to Germany. The entire marina was full, no one moved as the gale had been blowing for several days. The rigs howled loudly. On the evening of day three, the wind died down, and the sun shone through just before it set. I hear the woman, in the yacht next to us, reassuring her husband with the words: "It's over, it's over."

I have remembered and used her words many times since. They are so wise and useful for many situations. It is an art to be in the present, to not let the past overshadow the present. A situation we can't avoid anyway, so we might as well make the best of it. And notice that we can handle more than we think – even when it is tough.

The interesting thing was that those days in Gedser later became my best memory of the whole summer. Below is a reflection on the same topic I made at our passage of the Pacific Ocean. The crisis on my mind was the COVID-19.

This Too Will Pass

> Monday, April 6, 2020, 05°51'S, 121°53'W
> Marquesas Fracture Zone

Dear existence, how shall I handle this crisis?

> This too will pass.
> In the meantime –
> your work is to be present
> and take relevant actions according to
> what shows up in front of you.
>
> A crisis needs extraordinary actions.
> Like when you are on the sea,
> and suddenly it starts blowing more than expected
> – your relevant action is
> to adjust your sails to the new situation.

Stay alert.
Stay present and listen
– if you need to do anything,
you will know.

My willingness to accept change has increased a lot over the years. I adapt much faster and see an opportunity in the new situation. There is a reason that the saying: "what you resist, persist" is so often cited. Every change is an opportunity to show my willingness to accept whatever comes up. A perfect test of how mature I am.

How do you handle changes?

31

LETTING GO OF THE FEATHERS

"That you can't leave is your nature." – Osho

I was well aware that I had kept something on board, which to others have been easily thrown away, with the words, "kill your darlings". A friend in Stockholm, Lucie, even offered to perform a ceremony with me to release my darlings.

Throughout my adulthood, I had collected swan feathers with the dream of one day making my own wings. I had gotten to the point where I had chicken wire and plaster casts to create a frame for the wings. I knew that feathers were not allowed into New Zealand, the country in the world with the strictest entry/import regulations. I brought them anyway. I was not ready yet.

At Fakarava, an atoll in the Tuamotu Archipelago, and the place where Arthur had announced it was the fulfilment of his dream, I felt it was time to make the wings. I had the passage of the Pacific behind me. I had space now. I sorted out the many feathers and began to sew the long white swan feathers together into something wing-like. The only black one was a feather I got from a friend, Hamid, who said it was a steering spring. I always felt it was a great gift that I needed to take good care of.

Our Sacrifices

This was in July, and we knew that we had to sail to New Zealand in October, November at the latest. After that the cyclone season starts. The only problem was that New Zealand was no longer open due to COVID-19. It was very unclear if they would let anyone into the country at all. In fact, nothing around was open anymore. Tonga, on the way, was closed, so we also had to let go of that dream. Yet, we were determined to fulfil our plan A, to sail to New Zealand. I decided to use my wings as an offering and thank you, once we got our visas. It was the greatest gift I could think of, and I felt it was a fair exchange.

We sailed to Papeete and waited and waited and waited. Luckily, we had the safest innermost spot in the marina. If we had to stay, it was the best available choice for the cyclone season. But we continued to believe that New Zealand would let us in. As time went on, more and more fellow sailors changed their plans to other destinations. In the end, it was just us who patiently, well at least, waited for our visa.

On top of the feathers, we also invested money in an agent who wanted money regardless of whether we got a visa or not. And we had to promise to invest 50,000 NZD in refit and repair, a New Zealand requirement to allow any yacht, regardless of size, to enter the country.

And one day, on November 4th, we received an email from our agent, Duthie, that *Vista* had been granted permission to enter. The next step would be visas for us. It took five days before all the paperwork was processed. On the same day we received the final permit, we left Papeete for the Society Islands and Bora Bora. From there we made the final check-out on November 17th. The next day we were finally heading straight to New Zealand, and I let my wings out to sea. I was ready to fly high.

*

Only 40 yachts got permission to arrive in New Zealand this season compared to 400 in a normal year.

As you will see, the theme of letting go does not end. The feathers were not the last thing I could let go of. There was more to come, a never-ending story. Like the layers of an onion, we must start at the outermost.

What are your darlings right now?

PART III

OUR TIME IN NEW ZEALAND

32

CELEBRATING MILESTONES

"Having fun is all there is." – Arthur Sundqvist

Arthur and I have always celebrated the week. Usually, every Friday over a cocktail, when we share highlights from the past week. This time, we celebrated with friends at the legendary Duke Hotel in Russel, near Opua in the North Island.

Coming to New Zealand was a big milestone. It had taken us eighteen months of sailing to reach the other side of the world. We had once again fulfilled our intention and plan A. We soon forgot that we have had been sailing in a storm and had to tack against the wind for one of total three weeks to arrive in Opua. Arthur loves to tell the story of how he crawled out into the cockpit, and got horizontal waves all over him, to roll in our Genoa until there was only a small splinter left – while I slept.

Arthur and I love our celebrations. In fact, they are becoming more and more important. People tease us for being too rigid about our Fridays. Either way, it is something that keeps us aware of what is important in life, and what we want our main focus to be.

Do you have a certain ritual for celebrating?

33

MEETING THE KIWIS

"It all works out." – NZ saying

All the way from Stockholm, we had Auckland in sight. As big city dwellers, we thought it would be cool to live in the middle of their biggest city. In 2021, the year we arrived, New Zealand hosted the America's Cup – and they won and did so again in 2024 for the third time in a row. This is the America's Cup, and the United States have won the most since its beginning in 1851. In New Zealand, they have a strong contender. Despite many attempts to get a berth in one of the many marinas in Auckland, they refused us everywhere.

So, we ended up in Whangārei. A small town, also on the east coast of the North Island, two hours into the Hātea River. It is only possible to enter at high water (twice a day) and the very last thing required is a bridge opening. Then, after the bridge, there is a small marina, called Riverside Drive Marina (RDM).

We arrived at RDM the day before New Year's Eve. The other sailors met us, helped with the lines, and introduced us to the family rituals of a joint BBQ on Sundays.

As in many small towns, everyone took the time to greet and relate when we met.

The Kiwis

Kiwi is the nickname of the people of New Zealand. I thought it was after the kiwi fruit, which grows here, but no, it is after their national Kiwi bird (above). A medium-sized, slow bird that is active at nighttime. The Kiwis do not fly, but their feathers are much valued by the indigenous Māori people.

The Kiwi Way

We noticed a different pace – the Kiwi way – more relaxed and let us call it a bit more fluid with time commitments. In Sweden, we are getting things done on time, but here it usually takes longer. But sometimes also faster. They had a great sense of what was urgent, or not. Karl, the head of the marina, looked at us in amazement at this need to control everything. We wanted, for example, to determine the exact time and date for haul out (lifting the yacht on land). Karl knew how much could happen and affect the order – like wind, rain and current. He dismissed this European time-pattern and said with trust "We'll get there".

Softness

The surrounding landscape, or city, reflects the people who live there. New Zealand's North Island is green all year round. It is a subtropical climate and the hills are low. The surroundings feel soft. So did the people.

We were not only on the other side of the world; we were also in a small village close to nature. Very different from Stockholm. The slower speed would be very good for me. I got access to more of myself when I was in a new environment and meeting new people.

What kind of living space and place would be the opposite of what you live in today?

34

MAINTENANCE

"When you are in order, everything is in order." – Osho

Maintenance is life, the daily routine to keep everything in order. Like living on land, there are things we need to do daily and others that we do on a monthly or yearly basis. The difference on a boat is that we need to plan it much more.

We bring as many spare parts as we can; they are often specific, and we order them from Amel in France. Amel has an annexe facility outside of Europe, and that is in Le Marin, Martinique. We stopped there on the way over, and we already knew we would need to stop there on the way back.

Halfway around the world, in New Zealand, it was high time for a big maintenance job. All-important machines and their parts would need to be serviced. We ordered a new Genoa, and we had our standing rig replaced by Matthew. He had been a rigger on the *Steinlager*, a NZ Volvo Ocean racing yacht, for three years. It was lovely to see him up in the mast – he looked so happy.

And finally, thanks to the need for maintenance, we found a place in Auckland, at Oram Marina – right in the middle of the city. We went there to replace our battery charger. We were greeted with the sad news that the company had gone bankrupt. After many attempts, we managed to get our new charger. We had been able to borrow a generator from them during the order period, so that was probably what saved us from having to pay again. A neighbour sailor had to pay twice to get the autopilot he was waiting for.

A book about sailing around the world can be full of technical issues. Since there are already many of them out there, I will keep this chapter short, even though it takes a lot of time in reality. My focus is more on inner maintenance, which we will get to shortly. Still, it is important to keep the yacht, our home, in good condition because it affects us in so many ways.

How do you take care of your physical belongings?

35
TAKING PLACE

"I am not what happened to me, I am what I choose to become." – Carl Gustav Jung

Speaking of inner maintenance, something soon stirred up in me. At Riverside Drive Marina, it was a tradition to celebrate each other's birthdays. As my birthday approached, I felt the pressure to do as everyone else did. My resistance grew as the date approached. I did not want to be the centre of attention at all. I understood that I was hitting something unhealed in my history. It was not that I had not celebrated my birthday before, I often did. But this time I was on the other side of the world and barely knew anyone.

At this time, I was participating in a coach training program where I had weekly meetings with a group of coaches. I decided to bring up my topic to get some help. Perfect! I had a breakthrough and saw my history with clear eyes.

I decided to go all in and made a nice invitation card with a photo of myself as a child, I baked cakes, bought champagne and decorated the barbecue area with flowers, garlands and balloons. On my birthday, I started with a ten-minute speech, the first of my life! Earlier that day, I had received flowers from my brother and family. I was so touched by all the love I received.

After this episode, I saw how perfect a birthday is to celebrate each of us. Very fair and natural. We all have the right to take our place in the world. One of the very first personal development courses I presented had that very name: *Taking Your Place as a Leader in the World*. Even then, I was on the other side of the world, in Maui. I was shy and very quiet, just like when I was little. Writing books is now my way of taking my place in the world.

Is it easy for you to take your place?

36

UNEXPECTED HEALTH ISSUES

"Every day, in every way, I am getting better and better and better." –
Émile Coué

We, as people, also need maintenance. After a long passage, we needed to move our bodies, more than just up and down, to the cockpit. I also needed to get my hair cut, visit a dentist, and I was longing for a nice body massage.

Arthur could only walk a few hundred meters with crutches. X-rays showed severe osteoarthritis in both hips. We were in a small town on the other side of the world. In Sweden, hip surgery is free, but one must wait as there is a long queue. In times of COVID-19, we knew that this kind of operation was completely on hold. We decided to check out what was possible to do on-site. Fortunately for us, Whangārei had a small private hospital with nine surgeons. Arthur ended up having both of his hips replaced by them, one at a time. It took ten months for Arthur to have his first operation, and the second three months later. Against all odds, our insurance company paid for one of them.

This changed our itinerary. We thought we would stay in New Zealand during the cyclone season, December to May, and then continue west. It would be two and a half years before we continued.

COVID-19

We also got COVID-19, after a year and double vaccinations. For us, the pandemic was no big deal; we did not feel like it stopped us. The only thing was that it was tricky to travel. Not only were many borders closed, but the requirements for testing were also high. My passport expired during this time, and the closest Swedish embassy was in Australia. I had to have three medical check-ups, one per day during a short trip to Canberra. To our relief, the pandemic eased the restrictions when it was our time to head back to Europe.

※

We were not aware of either worn-out hips or a pandemic before we went. Thank goodness. If we were, we probably would not have made this life change. I see it as a good example of not worrying too much about the future. My experience is that we have what it takes to sort out the challenges that come our way. The key is to make your dreams come true while they are actual. Too many people wait and then get sick, or for other reasons, it's too late.

What is your experience with your ability to handle the unexpected?

37

WAITING FOR MY SOUL

"When I let go of what I am, I become what I might be." – Lao Tzu

Following our arrival in New Zealand, I had a strong longing to go on retreat and be quiet with myself. It may sound like a strange wish after a long time sailing with plenty of time on my own. I understood it as a need for integration of the big life change and journey we were making.

I had heard the story many times of the Indian, who got off the train and then sat and waited at the station. Someone asked him what he was waiting for, and he answered, "my soul".

That is how I felt, too. Everything had gone so fast; we had made so many changes in a short time. And not least, we had let go of a huge number of things, work, places, and habits and I missed my long-time friends.

I did not find a retreat, instead I found Tish, an osteopath. It was the first time for me to try this kind of treatment. I loved it, it was like healing.

In hindsight, I can see how this longer than expected time in New Zealand served both Arthur and me. I got plenty of time to go inward and complete my writing project. I was able to attend to my deeper and more subtle needs, those that required more silence and stillness.

Nurture My Female

It was also a time to nurture my feminine side. Life at sea is very tough and practical. Putting on a dress and even some perfume felt very nice for a change. I bought and picked flowers for the table and had long conversations with my girlfriends.

I need peace and quiet to hear my soul. A safe place where I do not have to watch over the outside, like in the ocean. This strong need came after a long time of being out at the huge and unknown oceans. My system wanted to retain the balance. As we stayed longer, we got more time for stillness than we had planned. That proved to be perfect as it opened up my inner universe.

When do you hear your soul?

38

THE GIFT IN SLOW ACTIVITIES

"Don't seek, don't search, don't ask, don't knock, don't demand – relax." – Osho

I have always considered myself as a fast person. I am quick to start and can go from words to action immediately. I would rather try it out for myself than read a detailed manual – much to my husband's dismay. Now, was the perfect time to meet new sides of myself. It was not easy to be as slow as Arthur wanted me to be, neither to be still nor wait for my soul. I had to face my impatience and frustration many, many times.

From Fast to Slow

In the West, so much is going fast, so I had almost forgotten the value of slow activities and stillness.

I had skimmed Daniel Kahneman's book, *Thinking Fast and Slow*. He explains that we get greater access to our brains when we are slow. That did not change my behaviour back then. Now, I had to. Arthur was super sensitive to speed and stress, first out of pain and later of rehabilitation. Our connection is the most precious we must take care of, so I had to adjust.

Imagine the following different ways of speed, from fast to slow.

Let us start with flying (about 1000 km/h). When we look down, we will see the big patterns of a city, a river, the major roads, the landscape, the coastline or the sea. Higher up, that too will disappear from our view, and we see the clouds and the sun instead.

Travelling by car is about ten times slower than an aeroplane. Now we can see the individual trees and houses. When we slow down in front of a crosswalk, we can even see the eyes of a child waiting to cross.

Sailing is a slower activity with an average speed of 6–7 knots, equivalent to 11–13 km/h, like running fast. Brisk walking is about 5 km/h. The interesting thing about sailing is the spaciousness. Even if it is a slow activity, we see the whole with the firmament and horizons around us, rather than details.

In the slow end we, for example, have the Zen monks who do walking meditation. They cherish awareness in every single step as well as what is happening in their minds.

To start listening for the soul requires silence and stillness. In my coaching practice, I have often invited my clients to close their eyes and listen inward for a while. Many reported that they heard nothing. And that is how it is in the beginning. We must practice this listening to open this channel.

My Way to Be Slower

It took me a while to see the value in being so slow, but the gifts kept coming, and now I am very happy with what happened. I can even read a manual if I have to.

My daily routine changed. I wrote a lot in my diary and in my BuJo, reflecting on what happened. Every afternoon I did a one-hour guided meditation, a perfect way to relax my body and soul.

Every day I took a slow walk and looked in wonder at the greenery and flowers. An abundance in this subtropical climate, where many species grow on each other. I always bring my iPhone with me to take pictures. Certain motifs, mostly flowers, caught my eye again and again.

Later, I found a beautiful book by John Diamond, *Beyond the Obvious – Photography for Healing*. He talks about photography as an action meditation, where he tunes in to the essence, the spirit. John begins his day with a lovely and powerful incantation:

"When I open my eyes I will see only Beauty, Blessedness, Belovedness everywhere."

I also swam with Leneke, and took as long strokes as possible to see how long I could glide in the water. On calm days, I let myself be rocked in the hammock. I read books, the thickest from start to finish in a day. We cooked slow food, for example, leg of lamb, and started fermenting sauerkraut. And with Arthur, I had long conversations where we listened deeply to each other.

Arthur started playing his guitar again. For twenty years it had hung on the wall. Finally, he got the peace to start training. It was very touching to hear him find his way back to songs that have been important to him.

Slow activities bring me to my being. A joking saying among meditators is:

"Don't just do something, sit there."

How about you, what is your natural rhythm?

39

WRITING MY FIRST BOOK

"A room without books is like a body without a soul." – Marcus Tullius Cicero

We were on anchor at our first atoll, Kauehi in the Tuamotu Archipelago – the paradise we had sailed long to reach. We had expected to meet turquoise water and snorkel. Instead, we got a gale, which blew hard for many days. We sat in the cockpit and looked back on the dinghy dancing by itself in the waves behind *Vista*. We did not dare to go ashore. It was then I felt, I wanted to write a book about my long-time coaching practice. From that day, I wrote daily. It took two years to complete and publish *Leading from Joy*. I felt like a pregnant mother passing over time before I gave birth to my first child.

When we finally went ashore, we learned that the name of the wind was *Maramu*. The same name as our yacht model!

Becoming an Author

Writing was another seed that began to grow. When I worked at Stockholm University for eight years, I wrote four research papers together with my colleague Johan. Our professor, Birgitta, recommended that people write a book instead of a doctoral thesis.

Later, I had a mentor, Kristofer, who suggested that I write a book while I was still actively working. The level of knowledge, writing a book would give me, would benefit my coaching as well. But writing a whole book didn't occur to me then, either. The closest I came was a few longer articles. Being an author was put on hold as a potential next career.

Finding the Structure

I was very happy when I figured out how I wanted to structure the book. I decided to build on what Professor David Hawkins had spent his life researching. In his most famous book, *Power vs Force*, he describes the map of consciousness. It starts with the lowest energy level of shame to the highest of enlightenment. He points out that a leader needs at least access to power, 200 on the energy level scale from 20 to 1000.

I found it a perfect backbone, as my coaching is very much about raising our awareness. I went back to my coaching and found cases that illustrated the different challenges. The final title of my book became *Leading from Joy – How to Transform 9 Inner Challenges*.

Writing in a New Language

Long before I knew I was going to write a book, I had decided to write in English. We have the world as our arena now, so English was an obvious choice. Not only because it is the only second language I know – it is also a very natural language in the coaching industry. That is how I learned to coach, and still, I am mixing Swedish and English when it comes to distinctions. Commitment for example.

I found support in writing in a foreign language in the Japanese author Haruki Murakami. He shares his experiences in finding the simplest possible words for others to understand him. "The cultural limitations and emotional baggage disappear in this process", he says. My experience is that a new language helps me be clear on what I want to say, and it also marks my new life. I appreciate learning a language that is one of the most spoken in the world, only Mandarin and Spanish have more native speakers.

New Lessons Learned

At first, I thought it would take a few months. I took a six-month writing course and learned more about the process. For the first time, I heard about editing – not only one round, but four, with different focuses. My idea so far, would have been for someone to check my English.

And what was my plan – to find a publisher or self-publish? That question took quite some time to sort out. Arthur suggested Amazon Kindle and encouraged me to self-publish. I read through the instructions, and felt it was impossible and very difficult – then.

Helping Angels

It was spooky perfect how I found my first editor. When the book was ready for editing, we had new neighbours at our pontoon at Riverside Drive Marina, in Whangārei. We introduced each other and I learned that Miranda was an editor and would be on shore for quite some time. She needed something to do while her captain fixed up their new yacht for the upcoming ocean voyage.

I used a second editor as well, and it was Rob from London. He was the first to respond to my request in a writing group on Facebook. Perfect. Both did a great job.

And later Robert, also an angel, helped me with both the cover and the layout. He also uploaded the first version on Kindle. He was very kind and patient. Now I can do this on my own, but back then it was super helpful.

Being Independent

Back to the publishing question. I looked at hybrid versions, where you pay someone to publish the book. This was before I found out that it is a big no, no. I spoke to a guy in London who offered that very service. But in the end, I did not get a good feeling from it. I decided to self-publish and become an independent writer. It turned out to be the perfect decision for me. I felt it was in line with my desire to be a leader in my life. It was better to learn the process myself than to hire someone for it. I named my publishing house Avalona Publishing and hired Amina to make a logo. I am still very happy with this decision.

I found writing to be a perfect way to complete this era of my working life. I had to rethink and reflect on what I had done, mostly intuitively. The book is the legacy of my coaching experiences.

What do you think would be your legacy from your experiences so far?

40

DARING TO BE EMPTY

"Be empty. Be still. Just watch. This is the way of Nature." – Laozi

After I published *Leading from Joy*, I felt empty. What had been with me for a long time, was no longer there. The book was out. Instead of marketing it, I felt a strong urge to be still and rest.

I walked a lot in nature. Now and then, I went through the coaching books and papers I still had and gave them away. I did not need them anymore. An era was complete. With the writing I had become clearer and this time I felt I had everything I needed inside me.

An Important Decision

I decided to leave the days blank for as long as I needed. The book was the completion of a long period of my life. I longed for space.

I also decided to wait and listen to myself for what I should do each day. As a results-oriented person who used to make long lists, this was quite new to me. It was perfect.

The passage across the Pacific gave me the outer silence. This time I was active in finding this space within myself.

This decision to listen and wait, instead of planning ahead, was bigger than I understood at this point. It was a watershed, a mark of something new that is still with me.

If you stop and listen for a moment right now, what do you hear?

41

LETTING GO OF CONTROL

"To let go does not mean to get rid of. To let go means to let be. When we let be with compassion, things come and go on their own." – Jack Kornfield

The longing to stay empty was a turning point for me. I was open to letting something bigger than myself guide me in life. Instead of thinking it all out, I made myself available to whatever the day had to offer. I let go of the old belief that I had to plan my life. What if I let my being and existence lead instead of my mind? What would happen then?

I changed my habits. I started the day by sitting silent in the cockpit with a cup of tea. Felt myself, heard the birds, looked out over the water. Curious and open to the new day. I got little messages; I wrote in my diary what came to me. After a while, I felt it was time to take the next step. I acted on what came to me, usually immediately. No more procrastination. No more stressing over things on the long list that I had not done. And so, the days unfolded, one step at a time. As our motto, wave by wave.

Signals

The signals from existence are often easy to recognise. Everything becomes so much easier and in a sense of flow when I hear and follow them.

A sunny and light, breezy day is a perfect day for washing. If it is very dry and hot, it is time to air and ventilate the lockers and the bed. A rainy day invites me to write, read and relax – the drops on the roof are so soothing. When it is cold, it is perfect to use the oven, it warms up a bit. The light wakes me up, the dark makes me sleepy. The body lets me know when it is time to eat, move or rest.

Then there are the signals that come from within, the hunches or the inklings, as we call them. I act on what is coming to me. I call or write the person I am thinking about. I try out a new idea. I take a closer look at what my eyes catch.

What I love most about waiting and listening to my soul is that I get the pieces that are missing. The ones I have forgotten.

The days often provide a lot of fluidity and synchronicity. I can get an inkling to do something that has been on my previous lists for months. Like a heavy burden. When I follow the way of existence, it goes easy and often faster than I thought. Each day ends with a feeling of being complete. I become much more present to what is happening, and I have all the time in the world. Perfect!

Letting go of control means accepting more of what can appear as chaos. If I do not judge, I often notice how perfect the divine order is.

My approach is to apply trust (no control) as an experiment. It is interesting to reflect on what I still want to control, and why. Like, for example, my spending and savings – as if that would make a difference. It could also be what we measure, like weight.

What do you want to control?

42

TAKING TIME TO REFLECT

"To be, or not to be, that is the question."
– William Shakespeare

After a few months of being and allowing emptiness into my life, a new idea appeared. Each evening, I would take the time to reflect more on the day that had passed. I would also choose a photo that represented the day. Since I usually take photos every day and had longed to make a picture book, I felt that this was a fitting project.

This was in early June 2022, and we had recently decided to stay in New Zealand. It was like getting an empty extra year. We were going to sail around the coast of New Zealand, that was all we knew. So, I decided to do a reflection project over a season; in this case, winter had started. It ended up being a year-long project, with the possibility of completing in time for our journey back to Europe.

I published these books as well, one season at a time. It was fun and rewarding to do something simple and short. Later I found a book by Patti Smith, "*A Book of Days*", where she also writes short notes to one photo a day for a year.

For me, it helped to highlight what I was present for and what was important to me in my life. The days did not just fly by. I stopped and learned to become more aware of details, patterns, and larger questions. I saw these reflection books as a documentary project of my development. It also answered my question: What will happen to me during a journey like this? The evening hour was also a good time to remind myself of what I was grateful for.

Everything Is a Mirror

Reflections are like the main road to our true being. The usual meaning of reflection is a re-mirroring. For example, when we see ourselves in a mirror, feel the warmth of a stone, or hear the echo in a valley. The whole world is a potential mirror for us.

How I experience what I encounter, see, hear or feel reflects my inner state. The energy in a meeting conveys a lot – do we trust each other or not? Are we being polite or trying to hide something? Is it fun or not? Usually, we project onto others what we cannot yet see in ourselves. All reflections are perfect material for knowing ourselves and becoming more aware.

Reflection time has become sacred to me. Whenever I feel lost, I sit down and start listening within, and then I usually write to sort out what is happening. By feeling what is important right now, I get to the essence. I also reflect on my photos. I look through them after the day's catch, to see what caught my attention today.

What is your highlight of today?

43

NATURE

"Nature is the source of all true knowledge."
– Leonardo da Vinci

The nature of New Zealand is magical. In the subtropical climate, everything seems to grow, all year round. The many versions of the fern tree brought me back to the sense of wonder. The fern is like a small umbrella tree, possible to stand under. The sprout symbolises a new beginning. I imagine it must have felt very special, and new, to reach this country so far away from everything else. The silver fern stands as a symbol of New Zealand's identity.

The nearby forest became a place I often returned to. It was such a nice contrast to the big city I lived in for most of my life. If the city was order, the forest was chaos. Plants grow on top of each other, and a plant could bloom and bear fruit at the same time. I also noticed that some of the flowers bloomed more than once a year – what an abundance!

Nature had its rhythm, even in this evergreen land. As most striking when I looked at the flowers. Nowhere else have I seen so many flowering trees in all colours from white, yellow, pink, red and even blue.

I met my first ones when we arrived at Opua. They were overflowing with red flowers and lining the harbour. Fascinated, I looked through my binoculars at this, so far, unfamiliar tree. I learned it was the endemic and famous Christmas trees, or Pōhutukawa.

Flowers

To my delight, I found many wildflowers. The first was Agapanthus, the wonderful blue or white big ball, sold at a high price in flower shops in Sweden. Here they grow in the ditches, free to pick! I felt like I was in heaven.

The next was a huge yellow something, with a wonderful honey scent. With my secateurs, I cut off a flower and took it home to *Vista*, as the trophy of the day. The sweet scent filled the whole salon. I found out that it was a wild ginger. In New Zealand, they categorise it as a weed, a garden escape, which the government is working hard to eradicate. After finding more and more wildflowers, I learned that most of them were weeds. I was happy anyway.

Uniqueness

Whangārei has a rose garden, a botanical house, the Quarry Gardens and many plantings in the small town. Like every garden owner, I was eager to see how the plants shifted and grew over the year. I loved to stay and look at the uniqueness of each flower. I felt more and more that each flower had its own energy and its own expression.

Bouquets

During our time in Whangārei, I always had flowers on the table and finally used my vases again. I loved picking a bouquet, once or twice a week. Often it expanded to several. I found wild, sweetly scented Freesia in the spring. The Camellias flowered all the year around and were perfect as a single one in a small glass. At the market, there was a man selling shrub flowers, such as Protea, the South African national flower. And a couple sold wild orchids in season. I loved the overflow of beauty.

Remember the old movie *Being There* with Peter Sellers as the gardener Mr. Chance? He could not read or write, and according to the story, he had not been outside the house where he was employed his entire adult life. When the owner dies, Mr. Chance comes out into the world and meets a wealthy couple, who mistake him for a man of wealth and deep wisdom. They introduce him to all their friends and say he is brilliant. Mr. Chance answers every question with his lessons from the garden, metaphors from nature. I loved that movie, and especially his garden answers.

Having lived in a big city most of my life, the close contact with nature is so nourishing. I feel like nature has an answer to almost everything of a more existential nature.

What is your connection to nature?

44

FILTERS TO PRIORITISE

"Excellence is never an accident. It is always the result of high intention, sincere effort, and intelligent execution; it represents the wise choice of many alternatives – choice, not chance, determines your destiny." – Aristotle

Very often, Arthur and I have morning talks, where we reflect on what is unfolding in our lives. Arthur shares his new findings. He is an excellent researcher, looking for the great souls. When he finds something interesting, he follows the thread back to the origin.

Arthur talks about the need for filters, and he is often mine. We can never take in everything that is there. We constantly have to choose, or at least that is a possibility.

One way to find filters is to search for the most skilled and passionate people in different genres. Those who have reached mastery, with high energy. Who has read the classic books that are still valid? Which authors can formulate today's topics and challenges? Which artists express the essence? What are the films, plays and books of the year? What and who is relevant in our time, for me and for you? Which companies are the leaders in different industries? What are their values? Who are the living masters?

When it comes to sailing around the world, Arthur uses Jimmy Cornell, among others, as a filter. Jimmy has sailed around the world many times in different types of yachts. His mission is to help people, like us, sail when and where it is safe. There are classic routes – those are the ones we follow.

Thankfully, there have always been explorers who seek and try new paths, like the Vikings, Columbus and other seafarers. The early ones, the inventors, the risk-takers and the pioneers with bold visions. We can learn a lot from those who have gone before us – those who have taken the time to understand the big picture, as well as the depth.

To Read and Listen

Even though we have plenty of time, we choose carefully what we read and listen to. In many marinas there is a place for book swapping. We see a lot of crime stories, but now and then we find nuggets of gold there too. We have replaced our large physical library with a digital one. Reading is an excellent and inexpensive way to learn from the masters all over the world.

The world has changed significantly in just the last century. The internet and the ease of travelling the world give us access to so much more than before. The options are endless. There is so much that is interesting. It makes me overwhelmed from time to time. I go with what is in front of me, often the simplest, instead of reflecting on what I really want and need. The grounding that farmers probably had are missing. The risk of indecision is that things are neither chopped nor ground, as a Swedish saying puts it.

This trip has made it clear that simplicity and a slower pace give me a more qualitative life. By doing less and being more, I experience life less fragmented and more whole.

How do you deal with all the information that surrounds you?

45

LETTING GO OF THE BATHTUB

> "There are two ways to be. One is at war with reality and the other is at peace." – Byron Katie

I had known for over a year that there was a bathtub in Smokehouse Bay. It is a popular cove for many sailors in the Great Barrier, a small group of islands off Auckland. There is also a pizza oven, a fish smoker and the opportunity to wash in large tubs. But it was the bath that attracted me.

Being able to bathe in a bathtub is the only thing I missed from our apartment in Stockholm. So, I had been longing and waiting and wanting to come to the Great Barrier for a long time.

Today is the day. We are on anchor in Kiwiriki Bay, so close that we can see that there is only one sailboat in Smokehouse Bay. We take the dinghy and see a family with children. I smell smoke and immediately start to worry that the bath is already taken.

We jump ashore and say hello. I quickly tell the woman that I am interested in the bath. She tells me that I have to fetch firewood myself up in the forest. I already know it will be very wet, after all the rain. She continued, "and then it takes three or four hours to get the water hot". It is already two o'clock, and I immediately realise that there will not be any bath today.

I go into the little house and look, and to console myself a little, I tell myself it does not look that remarkable. What did I expect, in the middle of the forest, on a small island? But that is not true, it is magical regardless to be able to bathe in such a place.

Then I take a short walk in the spectacular forest. It is slippery and steep, and firewood is not that easy to find, more like a project.

When I get back, Arthur talks to the woman. She comes from here and is kind enough to reveal where it is easiest to get fish, next to the Mussel Farms not far away. We say goodbye and go off to find mussels or oysters – they grow everywhere along the beaches. Most of them look small. The next secret we would like to know is where the big ones are, thanks. And tomorrow we will go fishing.

Back home, I rest in the aft cabin – the sun's rays come in and tease me. I feel like this is not going to work – I wanted a bath today. I decide to take a dip in the sea instead.

I bathe before Arthur. It has ever happened. He is the seal, (photo from Muros, Spain 2019). After all the years of Aqua Wellness, bodywork at 35°, I am more frozen than ever. But I lower the bathing ladder and jump in, without even checking the temperature. It is very nice, and I see that it is 22° afterwards.

I, who want to let go of as much as possible, realise that it is possible to let go of even a bathtub. I have a whole ocean to swim in!

The most exciting thing is that when we let go of how something must be, we always open up to something new. Thank you.

This text is also published in *Summer. Reflections and Photos from the Other Side of the Earth.*

Is it something you are longing for right now?

46

PAN-PAN AND MAYDAY

"Staying vulnerable is a risk we have to take if we want to experience connection." – Brené Brown

I keep coming back to the maritime safety communication, on VHF, channel 16. Mandatory to use for all seafarers, and to help if one is in the immediate area. The short conversations always touch me. In serious danger, one starts with Pan-Pan, which is repeated three times. If life is in danger, the distress call starts with Mayday three times.

I have clear instructions on how to do both of these conversations, that I read through in times of fear. I am relieved that it had stayed as an exercise.

Now I am going to present two real-life scenarios – there is so much to learn from them.

Pan-Pan

We are at Great Barrier in New Zealand when we hear the following Pan-Pan.

It is a man who is calling. Water is coming into his motorboat. He says there are two adults and two children on board. The coastal radio responds and repeats what they heard, "Is that correct?" Then they ask what kind of help do they need. The captain responds that he does not know yet. He wants them to know that he is in distress, that he never has been into a situation like this before.

Can you imagine how vulnerable the situation is? He is out with his family, and he understands that a water leak can be devastating if he cannot stop it. I could hear how scared he was. I could tell he had not made this type of call before (he was not following protocol). But he dared to tell the truth, he reached out early to someone who could help, if it were to become worse. He showed his vulnerability, instead of thinking he could fix it himself.

The rescue personnel were as always calm, respectful, guiding, repeating and asking questions. They followed up, until they knew he had stopped the leak and was able to make the trip home safely. They know that most people under stress cannot think clearly.

Vulnerability

This vulnerability and insecurity are part of being human.

"Forgive me, can you start over. Forgive me, I'm so nervous." Patti Smith had lost the words after the first line of *A Hard Rain's A Gonna Fall*. She was at the Nobel Prize ceremony to represent Bob Dylan, the Nobel Prize winner in literature, 2016. The audience applauded to give her the strength to continue, and she did so shortly after. She faced her vulnerability, at a very prestigious event, that was being followed by media from all over the world. It was an unusual openness, in a society where so much is primarily about looking cool and good. She received a lot of attention, and it became an act that many honoured Patti Smith for. The audience who was there saw her, felt her honesty and gave her support.

Mayday

Below is a mayday communication from Stena Scandica and Sweden Rescue (JRCC) on August 29, 2022. Stena is a large passenger ferry that runs between Sweden and Finland.

Stena: Sweden Rescue – Stena Scandica

JRCC: Stena Scandica – Sweden Rescue

Stena: "We have fire on board at deck four that we try to extinguish ourselves. No one injured right now." (They report their main problem, what they are doing about it, and whether lives are at risk.)

JRCC: The operator repeats what she has heard and then asks, "What is your intention?" (Because they have not asked for help yet.)

Stena: "We try to stop the fire ourselves. We are drifting between Gotska Sandön and Gotland." (He is still not asking for help.)

JRCC: "We have found your position. You try to stop the fire yourself." Then he asks, "How many persons are you on board?" (The main focus is always to save lives.)

Stena: "Crew 58, passengers 241." (He knows exactly.)

JRCC: "Well received. And you are trying to handle the fire yourself first, was that correct?" (Checking again.)

Stena: "Yes."

The JRCC then decided to send out a MAYDAY, as the ferry was not using its engines and could run aground. They continued to lead and guide a nearby ferry to get closer to Stena, and they also sent out helicopters. That was a wise decision. 32 people got evacuated, before Stena themselves stopped the fire and could continue. To be on the safe side, JRCC and a tugboat escorted the ferry all the way to Nynäshamn.

Even on this occasion, the captain was not sure whether he needed help or not. But he also called for help, because he was unsure.

*

The largest ferry accident in peacetime occurred in the Baltic Sea when the Estonia sank in 1994. 852 people died in the rough seas. 137 got rescued. The memorial is located right behind the Wasa Museum, near where we stayed our first winter with *Vista*.

Similarities on Land

The marine world can teach the land world a lot. The most obvious is to pay attention to the risks.

Water intake can result in flooding and drowning. In some part of the world, flooding is a real problem. For human, flooding can equal a situation of too much intake – of food, of work, of things to take care of, and so on. We say yes to be kind and good people. We fail to set healthy limits, and get overweight, overwhelmed and stressed. Stress makes us narrow-minded, and we make unwise decisions. The everyday business runs us, instead of the other way around. We know this is a huge problem. Do we call for help? My experience is that most often we continue to work hard, trying to solve it by ourselves as good girls and boys. Stress, depression and many lifestyle-related diseases are on the rise. If only we were more willing to ask for help at an early stage, I think we would have a better life.

Fire is always dangerous; it can end in an explosion. The comparison with humans is, for example, when we keep everything that feels heavy inside us. One day we will explode in rage. We say it is like walking on a minefield, not knowing when and where the explosion will come. All types of conflicts are urgent issues that cause a lot of pain for those involved. In its worst form, it is war with the intent to kill. The white flag shows giving up. In Sweden, many people are afraid of conflicts, so we keep silent and hope it will disappear by itself. Of course, it does not. Like rust on the motor, it will only get worse. Imagine focusing on safety first and helping each other.

Imagine that we walk around like traffic lights. The colour would change after our mood. Red and we are angry. Yellow, uncertain and green, happy, healthy and wealthy. Transparent and clear. Everyone would know and stop before the red one. Taking accurate actions to help. There is always a reason, both for fires and conflicts.

Lessons Learned

- Ask for help early.
- Respond, when someone asks for help.
- Check to see if you have understood the other person's intention.
- Be honest about the current situation and know your exact position.
- Know where and how to seek help before embarking on a risky journey.
- Have working communication and safety equipment and know how to use it.

It is no wonder that a VHF call often ends with, "Have a good watch", because safety always comes first.

As you can read, I am impressed and moved by the safety mindset at sea. The bottom line is saving lives. Quite the opposite of what happens on land here and there. In order not to be too overwhelmed by this, I do my best to show myself honest and kind to other people. I believe we have enormous lessons to learn about how we can help each other not only survive but also thrive in life.

Are you good at asking for help early on?

47

LETTING GO OF WEIGHT

"Beauty is the purgation of superfluities."
– Michelangelo

Arthur and I were not very good at asking for help. We were flooding ourselves with food. Good food. We loved it. It started already the first summer we met, in 1996. With romantic dinners, I gained five kilos right away. Over the years, our weight continued to climb. We made some serious attempts to fix it with month-long fasts on vegetable juices and such. Like many, we tried the 5:2 way of eating. It lasted a while, but soon we were back to the new normal again.

One day it happened. It was a sunny and calm day when we were on anchor in the Bay of Islands. Arthur said, "let's leave the excess weight behind us once and for all". He had plenty of time to think about how to find a lasting way to stay healthy. We were coastal sailing and everything was calm around us. The two new hips helped motivate him, and he read a lot of new research on health.

Arthur already knew a lot about health. He had read many books on the topic. To avoid white sugar and white flour with gluten, still applied. The new research, led by medical doctors from the US, could tell exactly how to turn the process around. When the body functions, it should burn fat and repair itself, instead of store the fat.

To make a long story short – we were both ready for this. So, we sorted out our cupboards of all the food that was not serving our bodies. From now on, it was only natural whole food that counted. And then we went on a strict one-meal-a-day, alcohol-free diet for months. The kilos came off quickly.

For me, it took a year to get back to the weight I had been at for most of my life, sixty kilos, instead of a most of eighty. Arthur has lost thirty kilos so far and is aiming for a few more.

This was a bold decision and the very best we could do for ourselves. We learned that major diseases as diabetes and Alzheimer's also are a result of bad eating habits. We did not want that, either.

Cravings and Co-dependency

When we stopped eating things our bodies were accustomed to, our bodies responded with cravings. The pattern is to want more and more of what we crave, to get the same effect. For example, when I had blood sugar drops in the afternoons, I could eat a whole chocolate bar to keep me going.

We can have the same needs of more: coffee, alcohol, sugar, drugs, gambling, risk-taking, shopping and sex. Even being busy, successful, rich and famous can be an addiction. We find out if it is a craving by asking ourselves: who are we if we are not feeding this need?

We are codependent (and craving) when something has to be a certain way for us to be happy. Anyone who has been around an alcoholic knows how much the mood can swing. Great, if there is alcohol in the body, but a disaster without it. To not make it worse, we adjust our own behaviour in an attempt to get the other person in a good mood again. The drama goes on and on.

It does not have to be that way. Gabor Maté, a Canadian doctor, has gone into depth on this topic if you want a reference.

New Food and Habits

Arthur understood early on that a big risk was that the whole food situation would become boring. We used to celebrate the week with a cocktail and have a dessert now and then. He studied new recipes and soon found out that it was possible to make very tasty and healthy desserts. For example, a chocolate mousse made with avocado. And what about our daily bread? Other flours, such as buckwheat and chickpea, were also possible. That was only the beginning of the new. Since that day, he is looking for new recipes that support our health.

We learned a lot of new things. For example, the need to go without food for a long time. When we do not eat, the body begins to clean and repair itself. The longer the fast, the better for the cleaners. The habit of snacking all the time is devastating.

We could skip many shelves in the stores. All those with jams, juices, cookies, pasta, white bread and many more. When you start looking at the ingredients, there is often sugar, bad fats and artificial additives.

I am fascinated by the slow food movement around the world, originating in Italy. These were places I wanted to visit. Now we have this quality on board.

Michelangelo finds beauty when he cuts away the superfluous. I find inner peace.

Support

To succeed with something as basic as new eating habits, we need the support of people who accept our choice. Especially in the beginning, when it is as the hardest to resist temptations such as sweets. It was perfect timing to be with ourselves and be out sailing during this time of transition to the new.

If you are curious about achieving sustainable health for yourself, check it out. The below recommended books are all written by doctors with extensive experience. They are also available on YouTube.

ketoCONTINUUM: Consistently Keto Diet for Life by Annette Bosworth M.D. (Dr Boz)

The Energy Paradox – What to do When Your Get–Up–And–Go has Got Up and Gone by Steven R. Gundry

Young Forever by Mark Hyman

It was incredibly nice to let go of the old way of eating. I was almost blind to my excess weight. I saw it in Arthur, but not in myself because I rarely look at myself in a full-frame mirror. The many failed attempts had made me give up.

I justified my rounded shape as normal, and that it made me look more feminine.

Afterwards, it is such a relief to no longer have afternoon cravings. It was also good for the upcoming long passages at sea. We both became calmer, which increased the security between us.

What are you craving?

48

CREATIVITY

"It's all in your listening." – Werner Erhard

We were both aware that anything involving creativity was part of our new life. I brought writing materials, and Arthur brought his guitar and watercolours.

I have always trusted my creativity, but now I have more time, and I get a lot of help from existence when I am out on the oceans. In the new environments, the familiar does not bother us. Four of my books started during a passage, and even my potential next project – a photo exhibition about light – came to me while I was sailing. Strangely enough, it has also happened in rough seas. What I write now is more of a waiting for the words. At sea, we have to be present and attentive to our surroundings, including our own state.

I think the water itself has a strong effect. In Stockholm, my most creative time was on Fridays, the day after I had led an Oceanic Aqua Balancing/Aqua Wellness class. In this relaxing bodywork in warm water, we tuned into the resonance in the body. When I gave Aqua Wellness, I held someone in my arms and gently rocked them above and under the surface. We get this slow movement 24/7 these days. That is calming, and all calm helps open our minds.

Letting go also supports creativity – it gives both physical and mental space for something new. At the beginning of our trip, I was so busy and amazed by all the new things I saw. I was busy understanding many things.

The more time passed, the more I could relax and just be. The challenge is not to immediately fill the void with the familiar, but to wait for the new from within.

The Captain's Cookbook

My dear husband has continued to experiment with new dishes. It started a long time ago in Stockholm when we decided to cook a new dish every Saturday. Our new way of life has increased the speed of embracing and choosing the new. New dishes are now often a daily surprise.

Arthur also has a book project in the works, *The Captain's Cookbook*. He will share all his new knowledge about food serving our bodies, not the other way around. It is already handwritten. Look out for it, he is a good chef.

I see creativity as a river. Sometimes filled to the brim with water, other times a little drier. Winding here and there in harmony with the surrounding nature. I do my best to follow the flow. I know when I need to sit back and be with the dryness. Other times I know it is better to take a break. And when it is rushing forward, I keep going as long as I can.

What creative projects are you playing with?

49

LETTING GO OF EMOTIONAL AND MENTAL STUFF

"It's just that feelings are only feelings; feelings are not facts and not who we are – and we can easily let them go." – Hale Dwoskin

Have you heard the analogy that our mind is like an operating system? Like in a computer or mobile phone. Or a new chart in the plotter, which needs to be changed when we make the big crossings. You know how often we need to update all the technical stuff. Very often nowadays. A ridiculous and true example of that is Microsoft (Word). They accept me working off-line thirty days exactly. After that, they do not let a single character come through on the screen, even though I paid for a year in advance. So, the technological world is a serious business – best to stay connected, logged in and updated.

The Mind

How quick are we to upgrade ourselves? Low consciousness is serious business too. Our mind is a tricky partner. Very sensitive to everything that is going on around us – how others respond to us, how the weather is, if something new is acceptable or not, all kinds of changes and so on.

We respond with stress or anger of all the input and choices. Or we may feel impatient, not accepting the way life folds out in front of us. Important trigger keys to shine some light on before letting them go.

And all that dependency stuff I just mentioned is run by hormones. Oh, we have amygdala, the brain part that oversees our emotional response. It would be good to have some maturity in that area! The tricky thing is that amygdala is a bit imbalanced and sees enemies in everything unknown. So, what do we do? Those are the hard things to let go of. Not at all as easy as upgrading an app, not yet at least.

Being together with my partner for many years requires kindness, respect and high awareness. Especially while sailing, it is a must. The challenge is not to get trapped in the devastating tricks of the human mind. I am not that fast, I notice, I hear myself repeating old ways of thinking and communicating. How many layers are there to peel off? My inner child keeps insisting on being loved, seen, and cared for. Being with my husband all the time seems to speed up the need to heal old wounds and memories.

Upgrading

Over the years, we have done many spiritual development courses together. It has been our way to reconnect and get more in tune with each other. We have got a lot of help to see patterns we were not aware of before. Meditation helps us get beyond the mind. The new way of eating helps our brains and bodies function better.

Imago Therapy

As we changed our eating habits, many emotions surfaced during the fasting process. It was time to tackle the next level of our mental and emotional baggage. As so often happens, we found an amazing resource at the perfect time. Sometimes I listen at Summits to see if I can find new and wise teachers. This time, the experienced couple Harville Hendricks and Helen LaKelly Hunt stood out. They have developed Imago Therapy, designed for couples. Both Arthur and I remembered that we had their book – a copy each. Neither of us had read it. Now we had plenty of time, so we decided to do their twelve-week program described in *Getting the Love You Want*.

Their genius is that they have understood the falling-in-love-process very well. On a deeper level, we attract a partner who is like our parents were – especially their negative traits. In doing so, our hope is to get from our partner what we did not get as a child from our parents. Have you noticed how similar your partner is to your father (or mother)? That is what they call the imago. Understanding this deep longing of getting the love we so much wanted, is the key. As partners, we can give this to each other now. In the end it is, of course, about taking everything back to ourselves.

Imago Dialogue

The core of their process is a special type of structured dialogue. The purpose is to heal the past and create a strong connection with each other. In the dialogue, I can take up everything that feels emotional and unclear for me. My partner will do his very best to understand my world by listening and mirroring what he hears. It is my responsibility to share my thoughts and feelings in a way that my partner gets them. In turn, he will show that he understands my world and perception. And finally, also show empathy by addressing the feeling.

It starts with asking: "I would like to have an Imago-dialogue – is now a good time?" In that way we show respect for the other, and when we say yes, we promise to be present.

Maybe you already can imagine that this is a time-consuming process. It is, and for a long time we did it daily. It took practice to accustom ourselves to the specifics of the structure. Being curious and having space to ask – "Is there anything more?" For the sixth time in a row. To understand the other to the point that he also feels understood, is easier said than done. Still, we feel it is very well-invested time. For us, it is a big change, and we will probably use the process for the rest of our lives. Our communication skills get better and better.

Other Ways to Let Go

There are many other methods that help us let go of what upsets us. Two more that we like, and use are Byron Katie and the Sedona method[*]. Byron Katie, ends her short process, the work[†], with the question: "Who would you be without that thought?" Used as an inquiry. When we see the pattern and how and when it began, we are usually ready to leave it behind.

---

[*] https://www.sedona.com
[†] https://thework.com/instruction-the-work-byron-katie/

What is your way of increasing your awareness?

50

LETTING GO OF THE OLD IDENTITY

"The first half of life is devoted to forming a healthy ego, the second half is going inward and letting go of it." – Carl Gustav Jung

I wrote a story about myself and published it on LinkedIn. My most read to date.

Once Upon a Time, There Was a Big-City Woman

Once upon a time, there was a big-city woman who was in the middle of her career. She loved the pulse of the big city and everything that was happening. She was dressed in black and wore shoes with sharp heels. She filled up with more and more stuff – in the cupboards, in the wardrobes, in the bookshelves, in the CD shelf, in the pantry and the fridge. She also added new exciting activities and new contacts to her network. Not only that, she also filled up the weight and depots in her own body and found lots of interesting mental influences. She worked as a coach and loved development in any way. There was always time to take in more of the latest in the field.

One day, she and her husband came up with the idea that they would sail around the world. Said and done, they sold everything and started a whole new life.

With a little perspective on her life, the big-city woman now realises that everything she used to fill up on was exactly what created both stress and excess weight. Because she did not have time to be present for everything, she thought would be good to have and know. It was just lying here and there waiting for better times. Now that she takes it a little easier, it comes up little by little, and she has the opportunity to reflect on its value and possible lessons. She sees in a new way.

Now she believes that life is more about taking away than adding. To remove what obscures the connection with her inner self, her uniqueness in this world. She becomes more and more curious about what her inner self has to tell her. She has understood that the whole point is to listen and express herself, rather than to adapt feverishly.

What are your reflections on my story above?

# PART IV

THE JOURNEY HOME

51

TIME TO MOVE AGAIN

"No pessimist ever discovered the secrets of the stars, or sailed to an uncharted land, or opened a new heaven to the human spirit." – Helen Keller

A big change lay ahead of us. It was time for us to leave New Zealand after two and a half years. This time to continue sailing – 21,000 nautical miles in foreign waters to Europe, with an estimated arrival one and a half years later. A long and risky journey with a great responsibility for the captain to bear. Like a mother carrying her child for nine months, but longer.

There is a tradition among sailors to wave goodbye and blow the horn, to wish the sailors fair winds and a safe journey. It is very sweet. We stayed for a long time in the small, friendly Riverside Drive Marina – a place and people we will forever remember with a warm feeling.

Route Choices

We heard several stories about the tricky Tasman Sea, outside Australia. Our neighbour had to abandon a yacht to sink out there. My captain decided to take the safer, more northern route via Nouméa, New Caledonia, and from there to Cairns, Australia.

We always choose the safest and most sailing-friendly way. For example, we skip Indonesia because there is very little wind in those waters.

Across the Indian Ocean, one can take the northern route, the middle route north of Madagascar, or the southern route south of Madagascar. Each with its advantages and disadvantages. We chose the southern route, passing Christmas Island, Cocos Keeling, Rodrigues, Mauritius and Reunion on the way.

Another choice, for those of us going to Europe, was whether to pass through the Suez Canal to the Mediterranean or to sail around Cape Horn in South Africa. We wanted to sail, so this was an easy choice.

The next choice is how to sail to Europe from the southern tip of Africa. There are many options. Some, in a hurry, cruise against the wind and the doldrums (passages with no wind), and head north. Others take their time to visit Brazil and other countries in South America. We took the classic route and went straight to the Caribbean, choosing Barbados as our first port of call. It is the most south-eastern island in the Caribbean. The South Atlantic also has a few small islands: St Helena, Ascension and Fernando do Noronha outside Brazil. We stopped at St Helena – the island where Napoleon ended his days.

And finally, where in the Caribbean do you depart for the Azores – the small group of Portuguese islands that are Europe's westernmost outpost. On that stretch, the Azores' high pressure meets the low pressure coming from the north. It creates a dynamic and tricky scene. Many take the direct and longer route from St Martin to Horta. We choose to go up north to Bermuda first, and then east to the Azores. This gives us a little more chance of setting off in favourable winds, and we also get to see beautiful Bermuda again.

For the route, I once again made our to-go plan with twenty-one destinations, twenty months and twenty-one boxes for every thousand nautical miles. Step by step, wave by wave, we would continue.

※

Everyone's journey is different and unique depending on the choices. We had the big picture clear of how we would sail back to Europe even before we left New Zealand. Even though the waters ahead were unknown, the destination was familiar, and we knew we loved it. We also felt we had something big left until the end, namely the Azores, which everyone spoke very warmly about.

How do you prepare when you have something big in front of you?

52

LISTENING FOR THE EXISTENCE

"It is only with the heart that one can see rightly; what is essential is invisible to the eye." – Antoine de Saint-Exupéry, *The Little Prince*

I take one day at a time and stay in wonder at existence. The birds circle and come to say hello. I take in the space and the vastness and notice the elements of nature floating in and out of each other. Everything is real and authentic here. The cloud is heavy with rain before it lets go. No shape is fixed, it is constantly changing. I feel grateful to be a part of this. We enjoy the moments as much as we can – celebrate small and large milestones.

On our most beautiful sailing yet, from Nouméa, New Caledonia to Darwin, Australia, I felt like I was receiving messages from nature. As usual, I witnessed the awakening of the morning. Short messages came to me that pointed to the wisdom of the present. I collected them together with a photo from that same morning in a little book, *Messages from Nature*. It is a beautiful poem, please take a look.

Many quantum scientists share the view that our imagination of the universe is perceived by nature, the collective energy field. That makes us co-creators of not only our destiny but also the destiny of the universe. *Messages from Nature* is my contribution.

The journey back home has begun, and we are greatly rewarded.

What do you feel grateful for right now?

53

THE UNIQUENESS

"Always remember that you are absolutely unique. Just like everyone else." – Margaret Mead

Every place and every person has their unique character. The beauty lies in this uniqueness. At Cocos Keeling, a small atoll in the middle of the vast Indian Ocean, there is only one place to anchor off a small piece of land. We stayed at Direction Island for ten days and got plenty of time to explore the details of the coconuts, a cable station, crabs and sea cucumbers.

Cocos Keeling used to export coconuts, so of course there are many coconut palms. There are different types of coconuts. A small one called marshmallow coconut, is a little fluffier and has a sweeter taste than the usual. Dried coconuts had fallen to the ground – if they still contained milk, they were edible.

Coconuts sprout easy. Imagine the soft green sprout finding its way out of the hard nutshell. That shell is not easy to open, we can tell you. Arthur first drills a hole to get the water out, and then he continues to drill a larger hole, finally using a knife. The locals use a machete or an axe.

A Cable Station

Behind all the palm trees we found a well-preserved history of a cable station. A heritage trail leads us around, telling the story sign after sign. This cable station was in operation until 1914 when the Germans destroyed it. It was rebuilt, but finally closed in 1966. Cables were the communication system of the time, with long cables on the bottom of the oceans connecting continents. Nowadays, this help comes from above, via satellites.

Crabs

We had just come from Christmas Island, where the crab migration was the largest in the world, so I was not surprised to find crabs here too. They were different and much smaller. Each one in a borrowed seashell, which protects them from being cooked in the strong sun on the coral beach. They too, ate the coconuts.

Sea Cucumber

The last speciality was a tip from a Chinese lady we met on Christmas Island – to look for sea cucumbers. They are considered a delicacy in China and are sold at a high price. We like to try new foods, so I was keen to see if there were any. And indeed, below the keel, I saw black shadows, like a cucumber in size, on the white sandy bottom. We picked some and did our best to cook them without a recipe. An Australian father and son, who were camping on the beach, snorted at our finds and said that it is only food for Chinese people. But the next day they were very curious about the taste. It was a bit like an octopus and needed to be boiled for a long time to soften.

When you live in a city and have a hectic life, studies like this can seem ridiculous. For me, it was reassuring to have plenty of time even for details like these. I see it as a life lesson, to discover the unique, not least in myself. Now it feels so easy – listen inward – hear, follow and express what is there. That is who I AM.

What are your unique expressions?

54

THE WIND

"Love is like the wind, you can't see it but you can feel it." – Nicholas Sparks

The wind is the invisible, feel-able and all-dimensional flow of air, influenced by solar radiation, that moves between high and low pressure. Wind carries heat, moisture and things over long distances: sand, pollen, seeds, pollution, ash and dust.

Wind provides the energy that propels us sailors forward. Its speed is measured in 0–12 Beaufort, 1–64 knots or 0–33 meters/second. We need at least a light breeze, 7 knots to sail, 14–16 is perfect.

Whether we are anchored or sailing, any wind strength from gale and up is a challenge. We have experienced them all.

Cyclone Seasons

A gale starts at 28 knots, a storm at 48 knots and above 64 it becomes a hurricane or cyclone, typhoon or tropical storm – they have different names around the world. The latter is always given a name and a category number, one to five, depending on strength. No insurance company will cover a boat that is in a named hurricane zone.

Atlantic cyclones start around Cape Verde and move westward toward the Caribbean. It starts in June when the water warms up. Many sailors head back to Europe at this time. Others have to stay put. Some places are said to be safe, like Trinidad and Grenada. As I write this, however, *Hurricane Beryl* has devastated Carriacou, near Grenada. *Beryl* built up to a category five when it finally hit the island on June 26 to July 2, 2024.

That is why we are so keen to sail during the cyclone-free seasons, usually six months at a time. In the Indian Ocean, this means sailing from May to November. If you want to continue and sail from east to west across the Atlantic, the cyclone-free season is the opposite, from November to May. It can either be from South Africa or elsewhere more north in Europe, usually Gran Canaria.

Gale in the Indian Ocean

In the Baltic Sea, even a gale is a sure sign to stay home. During this circumnavigation, we have sailed in gales many times. The toughest was between Cocos Keeling and Rodrigues, the easternmost small island belonging to Africa. It is a two-week sail in the most open and windiest part of the entire Indian Ocean, with waves that peak at five meters.

Storm at the Great Barrier

We encountered our first storm on the way to New Zealand. The next time we were anchored at the small island Great Barrier – riskier than being out in the open sea. Just a few days earlier, nearby Auckland had declared a state of emergency, as heavy rain caused the ground and houses to slide down. On top of this already urgent situation, a storm came, and the heavy rain continued.

We started to drag for the first time, the anchor could not handle the strong wind, so we had to move. We dragged twice more before we found a place at Kiwiriki Bay, with a clay bottom that allowed the anchor to dig deep. It was a small cove with a low hill right behind us. I remember looking up often to make sure we were at the same distance from the small mountain in the strong 180° turns. The water in the cove got muddy, so we could not use our water maker. If we had needed to, we could have used the rainwater – it filled the dinghy every day. It took a few tough and intense days before it was over. I felt elated after this experience; we had made it!

Cyclone Gabrielle

Not long after that storm, we had to experience the next level, a cyclone. They shall not go as far south as New Zealand in the summer. That is why people leave French Polynesia and Fiji during November to May, their cyclone season. But the unpredictable happens, we all know that.

We were heading north to the Bay of Islands when we first got the news of a potential cyclone approaching. The captain immediately wanted to take shelter in a marina. We had two choices, Opua where we were going – they were fully booked and Marsden Cove, a little further back. We called and luckily got their very last spot, and not long after the wind started howling. Although it was "only" a category three cyclone, it was the most devastating and deadly cyclone to hit New Zealand since 1968.

This was a story about the external weather we were sailing in. Another story is our own emotional weather. In Imago therapy, they have seen a pattern of opposite behaviours with the telling titles turtle and hailstorm. In our family, I am the turtle who keeps my emotions well-preserved inside me. Growing up as the firstborn made me a tough girl who managed a lot on my own. The good thing about that is that I am not easily worried even in rough conditions. The other side of the same coin is that I can appear too tough and emotionless. Sailing puts me in a context where everything is included and has its place. I learn and love it.

How is your emotional climate – is everything okay for you to express?

55

WAVES ≈ CREATIVITY

"Surrendering means surrendering to life; surrendering to the source from where you come and to where one day you will go back again. You are just like a wave in the ocean: You come out of the ocean, you go back to the ocean." – Osho

Waves show us the way. I am sure they love change. They are the perfect example of being in motion. They exist by moving. An individual wave only lives for a short time, building itself up, running forward – sometimes with a white plume on top – before reaching its peak and fading away – to then seamlessly transforms into a new wave, over and over again.

Waves are created by the wind. The more wind, distance and depth, the bigger the waves. The long-distance waves are called swells. In the deep ocean and in storms, extreme monster waves can build up and form a dangerous wall of water. A tsunami is a series of waves caused by an earthquake or an underwater volcanic eruption.

Close to the shore, tides and waves become visible – created by the interaction of gravity with the sun and the moon. The closer to land the waves get, the slower they will travel, and the top of the wave will reach its peak, just before it breaks. Even seemingly low waves breaking on the beach can be dangerous if we want to jump ashore from the dinghy. Contrary to the joy and delight of the windsurfers!

It is obvious that the waves are one with the ocean. They just change form – say yes to being both big and small, yes to all sizes and shapes. Such instant creativity! They just do it. Yes.

To be born and die, as if it were fun!

No judgement that up is better than down – both are needed equally. No looking back. Without hesitation they take a small turn and, if necessary, find a new way, faster than ever. And then letting go, surrender to the energy field, trusting that the new will unfold. Remembering that letting go is a prerequisite and makes room for a new creation.

Out here on the ocean, there is little to stop the waves – a small island here and there, a few ships and yachts now and then. Basically, they have a free ride for a long time. Like a big choir, they can take their time to build together from afar. Like Bolero, slowly building to a crescendo.

At latitude 40-50°S, the ocean is called the roaring forties, with strong and noisy westerly winds. We will stay far away – the closest we have been is at 35°S, at Cape Agulhas, the most southern tip of South Africa, where the warm Indian Ocean meets the colder Atlantic.

To sail gives a different feeling and rhythm than walking on earth. Out here, only existence rules. We accept and follow. We are connected. We also belong.

Lessons from the Waves

- We are all one, we belong together
- We need each other to move forward
- We are strong together
- You will feel more flow when you surf with us
- We know no boundaries – we make no distinction between you and me, yours or mine
- Without movement, we do not exist
- It is okay to peak! (it is gone in the next second)
- We gather strength in the valleys of the waves
- Life goes up and down, and that is how it is
- When we encounter obstacles, we easily find new paths
- We are experts in changing state and quickly reshaping ourselves
- The sea is our common playground

- Come and join us, we carry you too.

Our motto, wave by wave, has helped us slow down and be present to the wave we are on. Life changes quickly, and plans made far in advance often have to change. So, we have learned to listen to the waves within ourselves, as well as those on the outside.

What wave are you onto right now?

56

BE CURIOUS AT THE EDGES

"I have no special talent. I am only passionately curious." – Albert Einstein

In New Zealand there is a beach for everyone – big and small. We put wheels on the dinghy, aptly named *Beachmasters*. We explored many beaches. I was always drawn to the edges.

That is where the most interesting things happen – the unknown, not yet visible from the middle of the beach. At the edges, some kind of meeting of two elements, or cultures, takes place – like when the sand meets rocks. Edges act as turning points – either to stop at or to cross over.

As a committed leader, I have learned that I have access to everything. I am in touch with that mysterious part of life. In that light, it is even more interesting to reflect on the edges. What is still covered in veils?

A perfect place to get to know ourselves more is thus to seek the edges. What happens to us at the edge, and what opens up? Below I have listed some of the many edges we have around us. There are so many, I realised when I started thinking about it.

You may have heard the story of people standing very close to an elephant and everyone seeing different details, but no one seeing the big picture. Being curious about the many edges around us is a way to get a more nuanced view of the whole.

The Mental Limits

Let us start with our mentality, our inner edges. We cannot achieve what we cannot first think or believe. How big, or small, is our box? For many years, the Baltic Sea was my box. When I told Pia, another sailor that we could not get through the Kiel Canal because our engine was so weak, she just said "you can replace it with a stronger one". I blamed the engine without seeing that it was my mental block. Since I became an ocean sailor, I know for sure that I can replace my limiting thoughts.

How we talk to ourselves, and others, will inherently create boundaries. Check this one: What would be impossible for me to do, be or have? Ideas about my limitations emerges direct. Better to ask about what I long for. Or what will make me happy.

We can be superficial and deep. Our values are usually symbolised by an anchor, deep down. We say that we have to reach the bottom before we can make a U-turn. We can search for more, quantity, or go deeper for more quality.

As humans, we both have language and the ability to imagine new realities. The story we tell creates our lives. Is that a good one? If not, what would be an alternative? What changes do we want to see in the future? Changing the story of our life is possible!

The Sky and the Universe

When I was a kid, I loved the rhyme on where I lived – it started in my room and ended in the universe. My address never ended, as I belonged to the universe.

What happens up in the sky has always been of interest. With our eyes we can see birds soaring, stars falling and aeroplanes passing through and above the clouds. Birds and whales can cross the oceans and so can we, for about a hundred years. Neil Armstrong put his foot on the moon, in 1969. Next in line is Mars. More and more report that other living beings exist somewhere out there.

Wherever we are – when we look up, we encounter the spacious and transparent dimension of life. We can also imagine that we are zooming out until we see the whole earth as if we were an astronaut! We can even make imaginary time travel.

To see the whole and the big patterns, we need perspective. By taking a step back or climb a mountain peak – we come closer to the endless sky. How big is our movement pattern? Will we ever make it to the top?

There is so much to wonder about!

The Horizon

Let us come down to Earth. The horizon for the eye can be close or far away, like out on the oceans. I think of those who would follow Columbus beyond the horizon. They must have been afraid and wondered where they would fall when they reached the horizon – or the end of the world.

I feel freedom when I look at the horizon. All the space is equal to endlessness. The longer the view of the oceans, the wider my inner space becomes.

In Stockholm, my horizon was short – to the next house. My friend, Sati, shared the saying: "My thoughts can only go as far as my eyes can see." So true.

The Surface

The surface of the Earth can consist of water, soil, sand, rock, ice, snow, mud and I am sure much more. It will look differently above and under each of these materials. The temperature decides if it will be water or ice.

Below the water surface, we find ourselves in an enormous free and floating space. It is amazing to be able to swim and dive with the fishes! Submarines can go even deeper to explore more of the unknown.

In the soil, we sow seeds, and the roots grow long and reach the groundwater. Archaeologists dig for ancient finds. We say that we dig deeper to find ourselves. Our history and our ancestors are sometimes called our roots.

It is like the layers never end.

The Human

As humans, we also have a surface, our skin and manner. How aware are we about our inner life, and all the complex systems in the body? How deep do we breathe? Which speed is our pulse? If there is an imbalance, we see it on the skin.

How much do we show others? The Johari Window* is a visual framework that suggests we have an open area, blind area, hidden area, and an unknown area.

The Almost Invisible

Where is the edge of what we can see? When we zoom in, for example on Google Earth, islands will appear even in the very large and blue Pacific Ocean. With the help of a magnifying glass or a microscope, we can see things that the eyes themselves cannot see. Yet, the eye can only see a very small area of the light that is available. My iPhone picks up the blue light, in a way that my eyes cannot.

The third eye refers to our inner vision. We imagine something, we see a new image. And we can visualise both the future and memories from the past.

The Almost Unhearable

As we age, we lose the ability to hear certain high or low sounds. Animals hear tones that humans cannot hear. With the help of ultrasound machines, we can hear the heartbeat of an unborn baby or a whale swimming in the ocean.

I listen to my inner voice, to my thoughts and inklings, to the subtle messages from my soul. I listen to my heart and my breathing.

And when it comes to listening to each other, it seems like there is an endless amount to learn. Especially about what is said between the lines.

The Almost Intangible

I have a body meditation that I like a lot. It is a guided one-hour meditation. The voice guides me to trust my inner body wisdom. To listen to the messages, to the wisdom that I already have, but may not be aware of. I relax, I feel my body more and am more open to messages from my unconscious. I remember and can let go. I unwind from rigid patterns by dancing, or letting my body move as it wants.

---

* Invented by Joseph Luft and Harrington Ingham in the 1950s as a model for mapping personality awareness.

The Door

Let us come back to the materials. The door also acts as a boundary, to our home or someone else's. Inside there is the familiar, and outside the world to explore.

We close the door to gain privacy or open it to invite friends. We knock on someone else's door and wait for an answer, for permission, before we enter. The window also has the dual function of both protecting and opening up to let air in. Outside the house, the gate has a similar function. We open and close as we pass. We respect the symbolic boundary.

We may hesitate to open an unfamiliar door – we do not know what is behind it. In Sweden, we say that there may be "a corpse in the closet", meaning something unknown and unprocessed.

Sometimes we leave our homes and move somewhere else. We hand over the key to someone else, as we did on the day we sold the apartment and moved aboard. Others want to come back and thus keep the key, or as they say, "keep the door open".

An old and dilapidated house usually has its doorway intact. It is the safest place when cyclones blow. I find it exciting to step in – I step over the threshold with reverence – into another, for me, unknown history.

Even without a physical door, we can feel inside, or outside, a group of people. We want to belong and feel at home.

The Country Borders

Not everyone is allowed to cross all borders. It is not yet a human right to travel freely around the world. When we arrive to a new country, we have to follow certain rules – both to enter and to leave. We have to ask for permission, to get a visa, and higher authorities control us. Sometimes there is an in-between, a no-man's land, between two countries.

Border Crossing Aids

Traffic lights show the way and help us with the crossing. Here too, we are not allowed to do as we please. We have to stop and wait, let others pass and wait for our turn to cross. Bridges and tunnels are other aids to help us from one edge to the other.

We often need guidance when we embark on difficult journeys, when we must pass over troubled waters. In all myths, there are many forms of help available.

This became a long reflection. The more I wrote and re-wrote, the more came. I see all those edges, borders and perspectives as opportunities to explore something new.

Stay curious on the edges.

What edges are you aware about right now?

57
HOLY WATER

"The goal of life is to radiate happiness from each and every cell of your being." – Shiva

I bowed to the greenish water and took a handful of it and dripped it on my head. It was a sacred experience, as if I were being blessed. I took a few more handfuls – the whole feeling was very peaceful and profound.

We were at Ganga Talao, a sacred lake in Mauritius and one of the most important Hindu pilgrimage sites outside of India. A Hindu priest had a dream of a lake connected to the sacred Ganges River in India. In 1897, he found this lake and recognised it from his dream. Much later, in 1972, his dream came true when a priest came there with holy water from the Ganges. Never give up on a dream!

It was Nessen and his wife Audrey who took us there. It happened to be the day they were celebrating Ganesha, the elephant god, who brings happiness. We walked around the lake and came to the gigantic statues of Lord Shiva, and next to the goddess Durga, both thirty-three meters high. I was particularly drawn to Shiva. He is said to have removed all evil from the world.

We were given a bottle of holy water to take to *Vista* – for a safe journey back to Europe. With reverence, we have sprinkled the holy water over *Vista*, before all passages. As Shiva says,

"Those who believe. They receive."

All the islands we visited in the Indian Ocean were diverse in ethnicity, culture, language and belief. Hinduism was the largest religion in Mauritius, followed by Christianity and Islam.

I am one of those eclectics who say yes to what I feel is good for me and the world. I feel fortunate to live in a time when so much wisdom is available. The many quotes in this book are an expression of the world I live in.

What do you believe in?

58

SAILING ON THE OCEANS

"I [had] ambition not only to go farther than any man has been before me, but as far as I think it is possible for man to go." – James Cook

Sailing the oceans is a "yes" to travelling with existence. A "yes" to its waves, currents, tides, wind, doldrums, clouds, sun, moon and stars – parameters we can in no way control. The birds, the flying fishes and now and then whales and dolphins. These are the elements we relate to, depend on, and need to trust.

I had no experience whatsoever of the oceans before we left the Baltic Sea. The only thing I knew anything about, in theory, were the long and big waves. Below is my summary of the risks and opportunities of being out there. First a little introduction to the oceans.

The Three Big

The three major oceans are the Pacific, the Atlantic, and the Indian Ocean. The Pacific is the largest, covering a third of the Earth's surface. The seven seas refer to both the North and South Atlantic and Pacific, Indian Ocean as well as the Arctic and Southern Oceans.

The oceans cover a large part, 70%, of our Earth. They contribute a huge perspective we cannot get anywhere else. They are all interconnected and bring continents together. The salt water is our common denominator.

The oceans are enormous, compared to the smaller seas. This vastness creates opportunities for waves to travel long distances and build huge waves. Likewise, the wind gets free space to accelerate from one continent to another.

The oceans are unknown in many parts, very deep, mysterious and unpredictable, in constant motion. Uncontrollably. Only a fraction of the oceans has been explored. The invisible beneath the surface can represent our unconscious, and the element water, our emotions. The sea even has its own gods and goddesses.

The famous art museum, Louisiana, in Denmark, has Ocean as its major exhibition in 2024-2025*. In the introduction, they state poetry as the thread that ties everything about the ocean together.

Risks

At sea, we can face all kinds of challenges and extremes – the whole weather spectrum from calm to storms. From flat seas to high waves, with or without tides and/or currents. It is possible to encounter sandstorms, extreme heat or cold, thunder and lightning, rain, fog, hail, snow and icebergs.

We depend on everything on board to work – the yacht, its machinery, last but not least the captain and the crew. That is why every passage begins with a lot of preparation and ends with a lot of maintenance.

The long passages, say over a week, are always a safety risk. The further from land we are, the greater the risk – ultimately for life.

The price of a must-have insurance policy for the yacht says it all; it is around twenty times more expensive than home insurance. Comprehensive insurance is intended to cover situations where you have to abandon the boat.

The Most Dangerous

The most extreme emergency is having to abandon the yacht. The reasons can be leakage or problems with the rudder or the boat's propulsion. A hit by a whale (twice the size of a yacht), container or a ship can lead to water intake. Heavy weather makes everything riskier. Closer to land there is a risk of running aground.

Thunder and lightning are always scary. If we get hit, all the electronics

---

* https://louisiana.dk/en/exhibition/ocean/

can be knocked out and in the worst case, the yacht can start to burn. A fire in the engine room spreads quickly, and the yacht can explode – a catastrophe. Even a gas stove can catch fire and explode. We recently met a sailor where this had happened.

Even a cell phone can explode if it has a bad battery like Arthur's. Harvey, an electrician who was on board to fix some other stuff, happened to see that the phone was a little rounded. He told us about the explosion risk that we did not even know about until then.

Will the yacht withstand the forces of nature? The wind and waves can be so strong at times. The bow has to take the biggest hits. For this reason, Amel has made both the fore and aft cabins and the engine room watertight compartments, provided we close them properly. At sea, sometimes the wind comes from one direction and the waves from another. A yacht has to be able to withstand a capsize and spin a full circle, that is, right itself if it is pushed sideways in the sea.

Other Obstacles

Other obstacles will only make the trip more complicated and uncomfortable – all related to the propulsion of the yacht.

Many accidents happen in the marinas. The motor can stop functioning so one loses the capacity to steer. Twice we have been hit by other motorboats, once in a marina and once when we were at anchor. Both times in daylight and calm weather.

The propeller is a sensitive part, even though it is located directly behind the keel of an Amel. If a drifting fishing net gets caught in it, we cannot motorise when needed. To remove a net, one (Arthur) has to dive into the water and cut it off. Although that was our main argument for getting a diving certificate, it is not something we want to do, especially not out in the ocean.

Will the sails hold? Strong winds can wear out or tear the sails in parts. A little wind and big waves will make the sail flap – it wears a lot and is not good at all. We roll in early to avoid putting too much strain on them. Luckily, Amel has a very strong mast – a mast break has never been reported like on other yachts.

On long passages, sailing is a must. We have 600 litres of diesel – that is enough for five days of engine use, but it also has to be enough for daily generator use, mooring and emergencies. Many sailors say that no wind is harder to handle than a lot of wind, so it can be tempting to start the engine quickly. Not in our case.

Then we have all the machinery that needs to work. Besides the motor and generator, we have solar panels that provide us with electricity. Electricity keeps the autopilot, navigation system, navigation lights, radar, Iridium for weather and VHF running as well as the fridges.

We run the generator to charge our batteries, make fresh water and heat water. We have a large 1000-litre water tank. It has to be as full as possible because it is part of the yacht's weight – at sea, we want her to be as heavy as possible.

The bilge pump is one of the weakest points on the *Vista*. If it breaks, and it does from time to time, we must pump it by hand.

*

As you can see there are many things to prepare for. It would be devastating to think about the risks all the time. The Amel yacht is designed to make this type of trip a pleasure. How much fun it will be depends mainly on the mindset of the crew.

Risks with the Crew

We are the weaker part compared to *Vista*. Sailing solo or as a couple requires full fitness. We need a mindset where we support each other and trust our ability to cope with the passages. All forms of stress reduce our capacity to make wise decisions in difficult situations. Doubt and worry are devastating and take away the energy needed for the journey. Seasickness, fatigue, illness for any reason or having conflicts simply does not work. When we are in sync, it is calm – just like in nature.

We both have to be able to take shifts, especially at night. And in terrible weather, both may have to stay awake for a long time. We have an autopilot (and a backup) that steers *Vista*. They can break down – then sailing requires much more attention and energy. Captain is carrying a big load as the only one handling all the technical stuff.

The sun is so strong. The heat is so hot. Infections come easily, even in small scars. And the unpredictable?

As I said, man overboard is very dangerous. I spoke to a fisherman – and he said that it often happens when they pull in the nets. With just the two of us, it is a no-go zone. We always click on the safety harness when we go on deck. We know that the chances of saving each other on the open sea are very small. Falling is a risk – a big wave suddenly changes the rhythm when you are in the middle of a step. With a centre cockpit, we do not have the serious risk of getting hit in the head by the boom.

Food is extra important on a long trip. It keeps our energy up. Above all, there must be enough of it. We have to think about having enough canned or dried food in case the fridge and freezer stop working. And if the stove/gas stops working, we have to have canned food that we can eat right away. We also have to think about what is possible to cook on the open sea. We often cook all-in-one stews in our pressure cooker. And we bring extra bottled water, in case the water maker stops working.

*

With all these risks, it is always a welcome milestone to be halfway. After that, it is closer to continuing, than turning back. Imagine being on a passage that will take about a month, and something serious happens on day seven. Then you have to turn back, to the nearest land. It will change the whole journey. You may have to wait a whole season to get the right conditions to sail with the wind and current again.

Possibilities

Why on earth would we embark on such a journey of our own free will, you might wonder (unless you are a sailor)? At its best, sailing is a wonderful journey, in synchronicity with the trade winds and waves. A way to glide along in the flow of existence. All the risks mentioned can be seen as the ripples.

I believe the greatest possibility of sailing is that we must be aware, as we are exposed to powerful forces, without concrete walls to protect us. The continuous movement makes us feel alive and increases the grandeur of life. We learn to be present, here and now. We must be with what is – hiding, forgetting or postponing is not an option. It is scary to be in a storm, in lightning or out on a pitch-black sea filled with high waves. Staying with the fear brings great rewards. I feel blessed, confident, connected and more experienced; "We did it!" We celebrate often.

We learn to accept and relate to the natural changes of life. Our usual patterns are broken, such as sleeping through the night. We encounter new abilities that we normally do not use or are present to.

There is a lot of freedom in living and sailing in a yacht; like a snail, we have everything we need with us. And like birds and whales, we can move freely and over great distances. We also have each other all the time, which deepens our connection.

Sailing is an adventure, where we get to see new places around the world, some of which are only accessible by boat. It is fun to do something new and unexpected in life! There are plenty of opportunities to explore. My vision of what I can do in a lifetime is expanding.

Simplicity is a welcomed possibility. We have let go of so much. It frees up time and space. Emptiness is a prerequisite for making room for the new. We keep only the essentials, of good quality. We learn to prioritise what is most important.

The long passages in particular give us plenty of time to relax and just be. We can read books about complex and deep issues and have time to read uninterrupted and reflect on them. The distractions of city life and social media do not reach the sea.

Mirroring from the Ocean

The long time in the open and spacious oceans reflects and strengthens those sides within us. It is easier to remember the whole, the big picture and what is most important. The firmament reminds us of our vastness and to be humble before the unknown and existence itself. It connects us to the spiritual dimension of life. We often feel and express gratitude.

I love the sun and the light. I am light. Every morning, the sun rises. It seems so hopeful and trusting. Even when it looks dark to me, the light is here.

The light leads me towards beauty. The sun rises and sets and shows me endless variations of creativity. I love watching the sun and the clouds play together. No one takes over – they are all there, existing and expressing themselves at the same time.

We adapt to the rhythm of nature – we wake up with the sun, and when it is dark we fall asleep. We move at the speed of the wind, usually that means slow (5 knots ≈ 9.2 km/h). Other times it means being still and waiting for the wind to come.

Slowness awakens other parts of us – our ability to create, for example. But also hidden parts of ourselves, which finally see their chance to come out and get their fair share of our attention. Nature also reminds us of the bigger questions/rhythms of life, such as death, chaos and renewal.

When the world around us does not move as usual, our sense of balance reacts. Being in constant motion trains us to be in balance – physically, mentally and emotionally. Breathing helps us stay connected. Being in a clean environment helps us stay healthy and vital.

We also become more and more aware of the subtle sides – we notice small shifts in the wind and from the waves. We feel the wave coming early, and we hold on, instead of falling. We see things early and our intuition becomes stronger.

We learn from the surrounding nature. When the clouds are heavy with rain or pressure – they let go. The tension, the darkness and the fog always dissolve. Thunder and lightning are like great crescendos – with both sound and visual effects. It is ok to be very much now and then. The willingness to let go is crucial. We, too, can change states quickly if needed.

Understanding what it is to sail the oceans took time. This text came to me only after I had sailed the big three. Ocean sailing has given me a deep trust in existence.

What risks and opportunities do you see around you?

Read More at Noonsite

Below are articles I have written for noonsite.com about our ocean passages:
  Pacific Reflections
  Indian Ocean Crossing
  South Atlantic: Cape Town to St Helena
  South Atlantic: St Helena to Barbados

59
FLOWERS

"There are always flowers for those who want to see them." – Henri Matisse

I lay on the couch while sailing in the rough Indian Ocean, flipping through my photos – especially all the ones with flowers and bouquets. It comes to me that I should write my flower story.

For some reason, flowers have been a part of my life ever since I was a little girl. Back then, I would often pick wildflowers with my little sister. Depending on where I lived, my love for flowers has always found its form. During my time in New Zealand, this interest took up a lot of my attention. Studying nature and flowers helped me see my uniqueness and my way of expressing myself.

Where Flowers Grow, There I Go

Wander among flowers makes me happy and peaceful. Wherever I am, I try to find the botanical garden, the rose garden or the places in the wild where flowers like to grow. I seek the streets with houses that keep their gardens well-tended. In every shop or market, I take the time to at least take a look at their selection of flowers.

Bouquets

When we are in marinas, I always want to have fresh flowers on board. My collection of vases now totals twelve. The roses grown in South Africa were the best ever, they were so fresh and durable. For me, flowers represent beauty, something I love to surround myself with.

Floral Territorial

Did you know that the world's smallest floral territorial is at the Cape Peninsula? I got to learn that when we later came to Cape Town. At this southern tip of South Africa the species richness is so much higher than in the five other floral territories that exist in the world. Many of the most beloved cut flowers in Northern Europe come from Cape Peninsula. The largest floral territorial consists of the entire Northern Hemisphere, where trees are the common thread.

A flower is one of many metaphors for transformation. The seed already contains everything it needs to one day blossom into full bloom.

What is in your seed?

On Medium.com, you can read my article: *My Love for Flowers*
And on Amazon you find my flower story: *Beauty Everywhere – the Joy of Flowers and Inner Growth*

60

GETTING HELP OR NOT?

"Pause and remember – Every moment is a choice. Every thought, word and deed is creating your future. Choose wisely and positively!" – Jennifer Young

In the yachting world, it is quite common to use crew on longer passages. Mainly for safety reasons, if something were to happen, and you have to hand steer, for example. Others bring family and friends to have a shared experience. We decided early on not to have a crew, and that has remained the case. We know how fragile it is to be out there, and how important the relationship between us is. We love the passages and see them as sacred. There is a great risk that expectations will not be met, from both sides, with a crew that has not teamed up.

Weather Guidance

A common type of help during long passages is having someone on shore to help with weather routing. This is a person you have daily contact with and report your position and in return, receive weather, warnings and suggested route changes. We used it on the passage from Bora Bora to Opua. The captain was happy and felt supported.

For the Indian Ocean, we had not thought of it, and the captain trusted the weather files we got from Predict Wind. Anyway, in Darwin I got a tip about one that "everyone" was using. He did not even charge (that should have been a warning), so I signed up too. After all, the Indian Ocean is considered the most dangerous ocean to cross, and it could not hurt to have someone experienced with me, I thought.

How Dangerous Is It Really?

Every ocean is new the first time, so it is hard to get an exact feel for it until you have sailed it yourself. In total, it takes about three months to cross the Indian Ocean. I felt I needed to muster up the courage for it, in addition to believing and trusting in our ability to do it.

The good thing about the Indian Ocean is that there are small islands to stop at about every two weeks. The toughest stretch would be in the middle, between Cocos Keeling and Rodrigues – where everything is wide open. From other sailors, we had heard that there were high waves and a crossing sea. Others simply do not go there, because it is too risky – they would rather ship the yacht somewhere or sell it.

The trickiest part, we knew, would be the last one – from Reunion to Richards Bay in South Africa – 1400 NM, around ten to twelve days. First, we needed enough wind to sail to the southern tip of Madagascar. And there we needed a consistent current and wind to continue across to South Africa.

Between Madagascar and South Africa runs the strong and south-flowing Agulhas Current. Here too it blows strongly, but the wind changes direction every three or four days, from north to south or vice versa. It is a must to sail across when the wind is also blowing from the north. When the wind and current are at odds, monster waves can appear – the cause of the many shipwrecks off the coast of South Africa. Finding the right time to sail is not easy, as forecasts only predict the weather a week in advance.

A Decision We Came to Regret

Our weather guru was very reliable. We had our daily contact with short messages. As the days went by, the more anxious I became, and that in turn affected the captain. I got, in my opinion, overloaded with information about how dangerous it was. What was I supposed to do with that information, other than reef the sail? It was obvious that we needed to do it anyway.

There is a fine balance between courage and fear. Not being too brave and taking unnecessary risks. Nor falling into the trap of being too afraid, which limits our ability to make wise decisions, and not enjoying the journey.

What to Tell and Not

I remember when my father was dying. He was in a hospital, and we were all gathered around him. For me and my siblings, it was the first time we sat next to someone about to die, our one and only father. He was getting oxygen through a tube. Later I realised that the tube was gone.

The nurses knew that my father had already passed the point of no return. The oxygen was only there until we were all in place. The nurses did not tell us about that shift, because it would not change anything for us. It would be an example of information overload in an already fragile situation.

It Was Tricky

When we got to the southern tip of Madagascar, we got two pieces of bad news. First, one of the many things a captain worries about – we had a leak in the gearbox. Without oil in the gearbox, the engine would not work – a must to get into port. Second, the daily message from our weather router guy announced that severe conditions were ahead. He wanted us to either 1) head back to Madagascar (we were a day from land) or 2) head further north.

Since both options would involve more engine use, and both would extend the trip by many days, we felt we had to go straight over. That was also a bad option, but in our opinion the least bad. We informed the guy of our conditions and got the fateful answer: "I'm not sure if the gearbox is worth your life." With that heavy load in mind, we continued. Luckily, we survived.

But Not That Tricky

As if that were not enough, we received another worrying message from the "weather guru" before arrival:

> "In advance, a BIG welcome to SA – "Gangster's Paradise", so please make your personal safety your only priority. Trust no one – even some boats are suspect. R Bay is notorious for thieves, thugs and sharp operators."

You can imagine the feeling of arriving on this new continent where neither of us had been before. On my first little walk on the quay, I was very cautious and attentive. The first young man I met greeted me with "Good evening, Mum". I later learned that they called every older person they respected "Mum" or "Dad", even a stranger like me! It happened many more times. I felt so honoured and included.

The next person I met was a white-haired, smiling old man who introduced himself as Eric. He offered to take us to the yacht club that same day. And then came Natasha, from the South African volunteer organisation OSASA. She coordinated all the legal paperwork. Soon after, the health department came for a quick check for diseases and such. Everyone was very helpful and kind.

The next day, Eric took us around to customs, immigration, and to shop for food and buy SIM cards for Wi-Fi in a gigantic shopping mall. The legendary Zululand Yacht Club had a nice tradition of welcoming each new yacht with a bottle of local sparkling wine. A tradition we encountered in several other marinas in South Africa later. So it was not that bad. More like a very warm welcome.

Instead of thieves, we soon found out that the daily rate for maintenance work was very low. Many took advantage of the low rates and had their boats cleaned. But the sailmakers had prices higher than in the Caribbean. Eating out was also very cheap; braai – grilled meat or fish over an open fire, was their big thing.

As humans, we need a lot of help from each other. The question is with what and from whom. It is easier said than done to make wise choices.

I guess the most common mistake is the opposite – not knowing when it is time to ask for help. We are arrogant and think that we know and can fix things ourselves. We are blind to our incompetence.

Take personal development as an example – in the past, you only went to a therapist if you had to and were feeling mentally unwell. Nowadays, we know much more about the potential to mature and transform yourself over a lifetime. Does that make people eagerly seek this kind of help? Some yes, others do not believe that it can give them anything. We think we know, but in reality, we usually do not.

In the example above, I was unaware of my real need. If I had had a deeper connection with myself, I would have been able to communicate my need to be heard. Guessing and understanding what someone means beyond words is difficult. I understand that Arthur gave in, especially since I can be very self-assured. Following others, without being grounded in myself, is also an act of blindness. Being polite is a third. So, more awareness would have helped – guiding with the emotional weather rather than the weather outside.

The second question is choosing who to ask for help. What we ask for reflects our expertise in the field. Let us say we are beginners and, to continue with the example above, are about to choose a therapist for the first time. The questions we want answered before saying yes to a partnership are likely to be very different from the ones we have when we are experienced.

Part of being in integrity is being transparent about my knowledge when I say things are a certain way. What is my evidence for saying or doing this or that? What level of experience do I have? When I hire a professional, I want as honest a review as possible.

I know that I often have expectations, I believe a lot without checking reality. For example, it would be wonderful if every person was aware of how much emotions affect us/me. ☺

All of the above are reasons for me to continue my awareness practice.

What are your thoughts on getting help?

61

THE UNKNOWN SOUTH AFRICA

"Be where you are; otherwise you will miss your life." – Buddha

I have to admit we were both a little tense about coming to Africa. Our only experience of this continent so far was from the view of the Upper Rock of Gibraltar.

The only thing we had planned, apart from the sailing, was to order a new passport for Arthur. The only Swedish embassy is in Pretoria. That meant an overnight taxi ride to Durban, a flight to Johannesburg and then the local train. We had been told that white people should not be out after dark, and certainly not alone, so we went there together.

At the embassy, we got a pin with the South African and Swedish flags that we used a lot. Pretoria's nickname is Jacaranda City, due to the many Jacaranda trees that lined every street we passed. Until then, I had only seen a few trees, but here there were plenty – and luckily for us, they were in full bloom in lavender blue! Unfortunately, the rose garden was a distant memory, and the art museum had been closed for many years.

With that completed, we took on the next challenge of sailing down to Cape Town in the wild Agulhas Current. We were not particularly worried about this. Already in New Zealand, we met a guy from South Africa who told us the secret of how to eat this monster in small pieces. The trick is to set off directly when the wind blows from the north, in line with the current, and then to sail on and seek shelter before it turns south. With enough time, it would take us to Cape Town safely.

With the guidelines we had for the sailing it took us three weeks from Richards Bay down to Hout Bay. We stayed there for three weeks over Christmas and New Year, before we rounded the close by Cape and got to Cape Town. Below I comment on the places we stayed at.

Durban – a huge city we were also warned about. Even the black taxi drivers locked their doors when they passed through the city centre because of the risk of being robbed. For us, it became a very nice visit. A friend, Gordon, took us around and showed us some beautiful places in the area. They had for example a very well-kept botanical garden with large lotus flowers in bloom. Gordon was a white man who spoke Zulu, the largest indigenous language (out of twelve in total). His black nanny taught him from childhood. It was obvious what a good connection it gave him with the blacks.

At the yacht club, we heard the cool Kirsten Neuschäfer share her solo sailing around the world in the 2022 Golden Globe Race. She won after 235 days at sea. Three of sixteen boats (max eleven meters) finished non-stop; without any assistance, relying only on compass and sextant.

Port Elizabeth is a marina that people want to stay away from because it is dirty from the manganese industry. We only stayed overnight – there was really black dust everywhere. Industrial environments are often interesting. I got some nice pictures of herons hanging around the marina.

Mossel Bay is a place of contrast. After we managed to dock at the giant and rough stone quay in strong winds and currents, we found a lovely town. They had a cool café, the Blue Shed, which was also a coffee roastery. At a cosy restaurant, Kaai 4, with sand floor near the beach, they did the famous braai over open fire.

Here, human origins became real; in Mossel Bay, they lived! Archaeologists have found hand axes from the Stone Age, and I visited a cave where it was said that humans lived over a million years ago. *The Point of Human Origin* exhibition begins with the question: what is the most important thing we can learn from our early ancestors? Their answer is cooperation and sustainable living.

Hout Bay – a very windy and special place, is a fishing port surrounded by many mountains, where fires sometimes break out. I loved the powerful forces of nature. One of the delicacies they had was fish cheek. They also made large animal sculptures in metal and had art galleries.

The Cape, the southern tip of South Africa, is a World Heritage Site. Kirstenbosch National Botanical Garden displays the indigenous flora, including tree fossils 240 million years old. Their national flower, the Protea, is one of many fynbos species – the bush-like flowers that are very common in this area.

Cape Town – the most moving was that my mother sent me an old photo that my father had taken when he was in Cape Town in the 50s as a sailor and visited Table Mountain. From where we were in the city's marina, we could see it the whole time. One day we also took the cable car up, and it was a spectacular view and atmosphere.

*

If our entry into South Africa was risky, so was our exit. We visited one of the many vineyards in the Cape area and were on our way back to the marina. Everything had been excellent; the sun was shining, and Arthur was waiting in a big armchair while I took some pictures of the vines on the slopes nearby. Out of nowhere, a car speeds past him very close and fast – seconds later, he sees it crash into a car nearby. Shocked, he realised that he had been two metres from being run over, from death.

*

South Africa became our biggest surprise of the whole trip. It is a complex and dynamic country. There are problems, like high unemployment, especially for young people. And white people are having more and more difficulties in running their businesses.

On the other hand, the countryside is rich and the love for animals is palpable. Many sailors had the big five on their bucket list: the lion, elephant, buffalo, rhino and leopard. We had not even thought about it. When we finally went to the nearby Hluhuwe-iMfolozi Park, it was an amazing experience to see so many wild animals in their habitat.

Travelling is a great way to deal with prejudices – to see them, to let them go, to be more open to the complexity of a world/continent/country/person. South Africa turned out to be a good example – we got to see a lot we did not know about, even though three months is a very short time. We felt the roots of humanity.

Are you eager to explore the unknown?

62

ACCEPTANCE AND COMMITMENT

"Not until we are lost do we begin to find ourselves." – Henry David Thoreau

We were both very much looking forward to the long passage across the South Atlantic. The special thing about this ocean is the calm, windless passages, the so-called doldrums – a greater risk the closer to Africa. With the Caribbean and Barbados as our first port of call, we were hoping for pleasant sailing all the way. To our surprise, we got very little wind the closer we got to Brazil. The captain analysed the weather reports and concluded that if we did not motor, we would be stuck in this windless belt for weeks.

At this precarious point I added tension by making the serious mistake of not turning on the electricity first, when we were going to make water. The machine did not start. Arthur emailed Dessalator in Spain and through the dedicated technician he found out that the capacitor had broken. We had none in reserve. However, the technician knew a way to kick-start the engine by hand down in the engine room – with a shoe on his hand for protection. So Arthur had to do it for several weeks before he got hold of a new capacitor, the last one in Martinique. If the kick-start had not worked, we would have had to make a stop in South America, which we did not want at all.

All in all, it was a fantastic sailing anyway, and we are also glad we stopped in St Helena.

Many celebrations followed in a short time. First, we crossed the equator and thanked Neptune. Further north, in St Anne Bay, Martinique, we crossed our track, marking a completed circumnavigation of the globe. Shortly after, it was our 28th anniversary and a little later it was my birthday.

Coming to Martinique again was like coming home. It is the only place outside of France where Amel has an office, and a place where many Amel-owners come. *Vista* received much-needed maintenance, and, to my delight, a new washing machine.

Accepting What Is

It was Saturday night in Martinique. We had a nice dinner celebrating that everything was done. On Tuesday morning, the plan was to continue, a short trip to Antigua where their classic racing sailing weeks soon would begin. Could not be more perfect, we thought.

On Sunday morning, Arthur woke up with a high fever. He told me that his muscles had started to ache the day before. I suspected dengue fever because I had heard that the mosquitoes that spread it were here. On Monday morning, I also had a high fever, 40°. We stayed in bed and tried to rest in 30°, both inside and outside.

A Swedish doctor and Amel owner on site, Sten, told me that after four or five days of symptoms, one can take a blood test to check whether it is dengue fever or not. I did, Arthur was too tired to even move. The answer showed that I had dengue. Sten told me that there is nothing I could do to recover faster. The fever usually goes down after a week, it did for us too. The fatigue can last up to a month; it did. The risk is if your platelets are low, and you start bleeding – then you have to get a drip at the hospital. Mine were low, but I did not bleed.

This was of course a setback and the worst thing that ever happened to us. Instead of seeing some more beautiful places in the Caribbean, we experienced deep fatigue and, for me, severe dizziness. We knew we had to keep getting out of the Caribbean before hurricane season. Insurance companies require their insured boats to be outside a certain area at a certain time for the insurance to be valid. For us, that meant about a week of sailing northeast of Bermuda, in early June. We were both very shaky and weak and did not know how we were going to cope.

We decided to ask Ilkka if he would be our weather backup on the way to the Azores. Ilkka has sailed around the world three times, and we have met him and his wife Elina several times. The last time was some weeks earlier in St. Anne before they went back to Finland. They both said "yes" to be our remote contact for the weather. Thanks to them, we got that extra energy we needed to continue.

The two days to St. Martin started in thunder and rain. On the way, there was a problem with the generator on the newly serviced engine. The captain did not know what it was. In worst case, we would have to find a new one. Fortunately, the already mentioned excellent Harvey immediately took the time one Saturday evening to check things out. And to our relief, he was able to fix it. A new one did not exist, we had already asked. This is the yachting world! Things happen all the time, and they almost always get fixed, at least over time.

We were still tired, but Ilkka and Elena got us going, so we continued eight days to Bermuda. A lovely place we highly recommend. The final leg to the Azores in Europe took fifteen days. And we passed that insurance point, northeast of Bermuda, with only two days' delay.

To our delight, the last stop in the Azores, where we spent three months on seven of their nine islands, turned out to be a very good experience.

You're not home until you are. I think both Arthur and I thought that our time on the Atlantic and in the Caribbean would be easier and more filled with moments of enjoyment than it turned out to be. As sailors, we have a lot of practice in loving what is, especially with all the changes. Yet it is human to have expectations. And the lesson is to accept what is. Practice, practice, practice. For us, the backbone was our commitment to sail home safely.

Are there things you find difficult to accept?

63

THE LIGHT

> "There is always light.
> Only if we are brave enough to see it."
> – Amanda Gorman

I have developed an obsession with the sun. I am amazed by the sunrise. Morning after morning, I would sit and observe how the light grows stronger and stronger, until it bursts out and rises.

That moment of breakthrough marks the birth of a new day. This is the same given moment for every living thing on our planet. We can start anew, every day. We can take a step in the same direction, or in a new and different direction. We can choose to see the light, or we can snooze and sleep over.

The sudden sunlight, coming through a cloudy sky, is an opportunity for connection – as if the sun sends down its rays to meet us and show the path of Light.

Shakespeare calls the sun, "the eye of heaven". As if the sun sees us. I feel that it is true. It is as if light always comes at the right moment, sometimes just for seconds to give hope. And always after rain.

The Sun – Our Source of Life

The sun is our biggest source of energy. Without the sun, there would be no life on Earth. The sun is also our main source of vitamin D, a hormone that supports approximately 2000 genes in our bodies.

Remember I stated that it starts with falling? Arthur has found a scientific answer to why falls and osteoporosis are so common in Sweden. According to MD Michael F. Holick in *"The Vitamin D Solution"*, it happens because we, in the north, do not get enough sunlight. For example, it is twice as common to develop multiple sclerosis compared to those who live in warmer countries. We all need a small daily dose of sun, and the skeleton will become stronger.

There Is Light and Beauty Everywhere

Our saying that "the sun rises and sets" belongs to an old paradigm. Even with the knowledge that the Earth revolves around the sun, it is still abstract. What I see is not the whole truth. The sun is there all the time. The light is always on. The sun is always visible somewhere, and it can be seen above the clouds.

I scroll through my pictures, wanting to find a very beautiful one to show you. It is difficult, because I think they are all amazing. Then I realise that the most beautiful sunrise is always the one I see right now.

The same goes for people – the most precious moment and encounter is the one I have right now. Imagine that light within us humans – it is there – but sometimes we cannot, or do not want to show it, and from the other point of view we may not see it. With a smile, the light within all of us will become visible.

Imagine if we all also knew how to turn on the light and make it brighter. Starting a little cautiously in the grey and then playing in a little yellow and pink. Slowly amplifying, intensifying ten times more, using the whole spectrum of ourselves. Shining with pure joy, with beating hearts. We bubble and play to the max, a crescendo, before we retreat to rest. Happy in every part of our bodies. Every part of us has been involved in its own way. Like a big theatre ensemble playing together. Just for fun.

Do you know how to make your light brighter?

64

THE BIRD

"Man is condemned to be free." – Jean-Paul Sartre

Now and then, seabirds or pigeons land on *Vista* while we are sailing. They are very welcome. They rest for a few hours before continuing. Usually they are shy – they sit at the stern and quickly fly away if we move in the cockpit. Others are content to fly around us at high altitude to connect. I feel them even though I am inside, so I go out and look and say hello. They make me happy.

Now we have also had a little songbird visit. The nearest land was Santa Maria, our last island to visit in the Azores, which we left about thirty hours ago. Maybe it is the endemic bird Estrelinha that we saw on a large mural in Vila do Porto? Either way, we feel very honoured by this unusual visit.

She is different. She flies into the salon and then to the forepeak. Checking it all out, sitting here and there – on the dangling onion net, on the lamps and even on us! And then she flies up to the cockpit – strutting around the instruments – and then she comes back down again.

Arthur gives her some water in the sink, and I put out some sesame and hemp seeds, which she rejects. Instead, I watch her catch a bug on the curtain.

Neither Arthur nor I have ever had a bird come this close. She was completely fearless. Arthur reached out his hand – she came and sat on it, and later on his head.

The energy she brought was very sweet and tender. Even though we saw how confident she was, we spoke lower and moved with grace to match her energy. What a great teacher. No words were needed.

Synchronicity

Arthur felt the bird was carrying a message from a new angel named Beatrice. At the same time, I was listening to a podcast with Beatrice Harrison, known as Lady Nightingale. She played cello, when a nightingale in her garden started to sing along with her. Later they recorded it and that was how she got her nickname.

I love all kinds of birds. They are so obviously free – flying around in the sky, even more spacious than the sea. They are so different and unique – imagine an eagle next to a nightingale! But both fly above us. They must see very well. And I imagine their internal GPS is very reliable and brings them home when it is time to nest.

What reminds you of your freedom?

65

THE MOUNTAINS

"It is not the mountain we conquer, but ourselves." – George Mallory

The birds and the mountains are our signs of land. Visible from afar. It may be hours before we make the landfall. Arthur looks enthusiastically through his binoculars at the distant landmarks.

The mountains are our protectors. It is behind them that we seek shelter from the wind and waves. We enter as far as we can into its bay. A high and straight cliff provides the best protection. A valley can invite down winds.

The mountains just stand there, like the essence of all being. They witness both beauty and chaos, both light and darkness, without moving or judging. They accept what is with dignity and love.

They are old, many millions of years. When large continental plates collide or when a volcano spews lava, a new mountain is created. Like the annual rings of a tree, the layers of rock and soil tell us about their different ages. They have seen much more than any living creature.

They give us a higher perspective. Climbing to the top is almost a human need. They let us climb them. At the top we will see more, the other peaks and the larger context. We can be the watcher of the hill. At the bottom of the mountain, we experience how small we are. The first step seems to begin right there.

Having lived in a big city with houses all around me, I was rarely in places that gave me perspective. Now, after a long time in pure nature, I see the value of having contact with as many aspects of nature as possible. Otherwise, life will become too unbalanced and narrow-minded. The birds fly freely in the sky. The mountains are deeply rooted and do not move at all. We need both.

I have long carried a deep feeling of the man symbolised as the mountain, and the woman as the storm/the ever-changing weather. There is even a saying – he is the rock in my life. Now, I believe that this ability to be a rock is something we all need to get in touch with.

Who or what is the rock in your life?

66

LETTING GO OF THE PROFESSIONAL TITLE

"Drop the idea of becoming someone, because you are already a masterpiece. You cannot be improved. You have only to come to it, to know it, to realize it." – Osho

What title do we write on the business card? My husband and I have a card for *Vista*. In the sailing community, we often refer to each other by the name of the yacht, rather than by personal name.

ICF MCC

On my business card, I have referred to my master certification as a coach, ICF MCC, for many years. ICF stands for the International Coaching Federation, the oldest and largest professional organisation for coaches in the world.

I have a long history with them. I became a member in 2002, acted as a chapter host in Stockholm in 2005-06 and certified myself step by step. First to become ACC, Associated Certified Coach (2006), then PCC, Professional Certified Coach (2009) and finally to take the step to MCC, Master Certified Coach (2015). When I got it, only 3% of coaches in the world had qualified for that level, 4% today. I was also on the ethics committee.

Every three years we have to renew our certification and prove that we are active and continuously educating ourselves. My next expiration date was

December 31, 2024. I thought a lot about whether to renew again or not. So much has changed in my life. My work life will never be the same – full-time with a physical office in Stockholm. I decided to end it. The question I still wondered was who would I be without that title?

Lightening Your Load

I got the perfect help – again. Dr. David Drake announced a six-week program called *Lightening Your Load*. Drake has been in the industry for a very long time, but is not part of the ICF, as he advocates, among other things, that maturity is much more than what is represented in the ICF core competencies. I thought this was a good opportunity for me to get more in touch with his teaching. This program was based on his recent research on The Five Maturities.

During the program, we were invited to bring an important question, to think about over time. I had the perfect question for this. I am very happy with the support I received. I got a safe place for reflection – and the answer came.

Being Independent

At the end of November 2024, I came up with my new title for LinkedIn. Here you can read why I chose to call myself Independent.

☼

We will all come to a point where our professional lives end. Our profession is often a big part of our identity. Who are we without it? Even if you, like me and many others today, are doing a part-time downsizing as a transition to something new. The question remains – who are we without our professional title? I postponed the question by adding a new profession, as a writer.

The bigger question is who are we behind all the roles we take on in life?

67

SAILING HOME

"Taste the Ocean from anywhere, and you will always find it salty." – Buddha

The big movements in the outer, over the oceans, have stopped. On the early morning of Tuesday, 26th of November, we arrived in Las Palmas. Two days after the big armada of ARC-cruisers left for the Caribbean. We thought it should be easy to get a place then, as more than two hundred boats had left. That was not the case. There were still many yachts out on the temporary anchor place.

We had gone to the marina pontoon in the hopes of entering. The marina guy told us to leave and come back in the dinghy and ask to be placed in the queue. What a complicated system! Taking down the well-secured dinghy, and back up again as we wanted to come into the marina, was a couple of hours job in total. The guy saw I was starting to get irritated. Arthur was the kind one who showed understanding from the marina guy's perspective, but also told him that we were just following their email-instructions. How it was, the marinero made an exception and was kind enough to put us in the queue and asked us to call back in two hours.

We went back out to the anchorage, ready for a longer wait. After two hours we called, this time without any expectations. To our surprise, they told us to come right away. A dinghy from the marina showed us the way to the very best spot in the marina – by the pontoon for the big yachts. We had not asked for it, but it was the place Arthur had secretly wanted! Mission accomplished; Vista is now safely moored for a longer period of stillness.

The Port

The port is a place for both departure and arrival. It supports us, acting as a nourishing cradle – giving us all the shelter and resources we need. From the same place, the sailors both start and finish their great adventures. Side by side, each yacht with its own story.

I recognised the feeling of excitement and nervousness of the crew next to us in the first days – six guys who, with loud voices, made their final preparations. They will also become calmer and quieter after having been out on the ocean for weeks.

Integration

I, once again, had a strong longing for being still. I remember how tired I was when we got to New Zealand and then understood it as a need for integration of all the new we had experienced. It was the same situation now, with the exception that I was not physically tired. Only a deep desire to integrate the journey of sailing home, as we see the Canaries as our new home-place on earth.

For the first time in almost six years, we were in Europe with easy access to fly home and meet family and friends again.

Arrival To-Do's

We have a ritual when we arrive and know we will be staying for a longer time. It always starts with cleaning, inside and outside. I am in charge of washing the deck, and Arthur does other things like taking down the instruments and checking the machine room.

Only now do I unpack our grab bag. It is a great relief because it is a tangible reminder of the dangers of life. It is where the most important things would be if we had to abandon *Vista* and jump into the life raft. You do not have a grab bag under normal circumstances.

We unplug the freezer and stop stocking up since fresh food is within reach every day. We end the Iridium Go subscription which gives us sea weather (€2800/year) and buy local internet.

Step by step, I go through everything on board. What we no longer need, I give away or throw away. Gas bottles from New Zealand, clothes, books, etc. Then it is time to get other things out again. Fragile ones that we never use while sailing, like ceramic bowls, mixer, electric stove, candles and most importantly vases for fresh flowers.

On a deeper level, we began to process the homecoming already when we left New Zealand. An example of this is the literature we read and what I write. Back in February, during the passage of the South Atlantic, I wrote the tale of coming home, based on the Major Arcana (15th-century wisdom of life). It begins with the courage to jump and begin a new journey. The following twenty-one steps, until the journey is complete, address all the twists and turns that happen along the way.

Later, Arthur found the book, *Sailing Home,* by Norman Foster, about Odysseus' long and dramatic journey home. This ancient Greek epic was written around the 8th century BC and is still applicable today. The human challenges and how we deal with them have always been of interest. I saw the similarities in our journey and appreciated the sailing details. There are true heroes and heroines in every time and part of the world.

If you think about your life right now — where would you say you are on the journey? At the beginning, the middle or the end?

PART V
REFLECTIONS

68

TO MAKE IT COMPLETE

"A man travels the world over in search of what he needs and returns home to find it." – George Moore

I met a sailor who had noticed that many sailors were ending their trips around the world earlier than planned. It often happened in New Zealand, or Papeete – with the beautiful Pacific Ocean behind them, and the tougher journey to Australia and South Africa ahead. The three main reasons were 1) running out of money, 2) declining health, or 3) the relationship wasn't working out.

Our biggest worry before we left was whether we would have enough money. Even before we left Stockholm, we had setbacks in that area. But because we had committed ourselves, we stuck to our plan and trusted that it would work. And it did, thanks to coaching clients. All the other challenges with COVID-19, health problems, accidents and of course tough weather and sailing worked out too. We did not let any of that stop us. Our health is better than ever, and our relationship has deepened.

The feeling of having succeeded is amazing. I am so glad I listened to that little voice that suggested a life change. I have experienced and learned so much. In advance, I could only imagine a fraction of the journey and the gifts it would bring me.

The Outer Journey

One definition of complete is to refer to the outer journey. My world has truly expanded, and I have my own sense of the size of our globe. We have visited new continents, not just countries. Places we had only read about before; we have now sailed to on our own keel. I have my own experience of those exotic places, and especially the big blue abstract Pacific Ocean that covers half the earth.

It is a wonderful feeling to be able to say, we have sailed around the world! I am grateful to have a skilled partner who is brave enough to be the captain of an adventure like this.

We have carried out our plan A, to sail with the current and wind at the right time of year. It took us five years to get back to Las Palmas.

The major geographical milestones were the Panama Canal, New Zealand and Cape Town. On the way back, the Azores was the most important milestone.

Together, Arthur and I have made it through the tough route across the Indian Ocean and around South Africa. A stretch that many are still hesitant to take – and for good reason. We did it!

Fun Facts

A French team has so far done the fastest circumnavigation of the globe to date, non-stop in forty-one days, in 2017. And the World ARC Rally, which runs the same route as us, usually does it in about fifteen months.

The first circumnavigation of the globe took place in 1522 by Ferdinand Magellan. He was a Portuguese explorer who discovered the Strait of Magellan in southern Chile. At that time, the Panama Canal did not exist. It was not until 1914 that the passage was possible, although the idea of creating it began as early as 1534.

The Inner Journey

Then there is my inner journey. I have plenty of time to face myself, my fears and my gifts. I am in touch with my being, and I listen to her as much as I can. I have found great gifts in being slow, and I have learned to accept endless changes. They always come with beautiful gifts. Now I feel that I have access to both my inner man and woman, both my intellectual and creative parts. I trust existence and share its beauty. We both feel blessed and grateful for our new way of life.

The biggest thing is the hardest to explain – it is a feeling of being one with the universe. On a sailboat, the forces of nature are so tangible – we have to adapt to the wind, waves, currents, tides and much more. We feel that we, all here on Earth, are part of such a larger and infinite planetary system.

That is something to celebrate!

Being a Team – Being Ocean Sailors

Now both Arthur and I know what it is like to sail the oceans. We have fulfilled the intentions we set five years ago. Arthur still does most of the sailing and navigation and is the engine room manager. We have found our way of sailing together, where I still have plenty of space to write and read.

We have experienced that the small team of two of us has everything we need within us. Together we succeed in this adventure. Of course, I take shifts too, and I learn more and more as I go. We are partners, and now I playfully call myself admiral.

Sailing the oceans has led to increased awareness and presence. We help by spending a lot of time letting go of what belongs to our minds and our history. The more we both stay alert, awake, aware – connected to ourselves and each other – the greater the chance that things will go well. Life does not have to teach us very tough lessons then. We have already taken care of and adapted it well in advance by listening to our hunches, and our intuition. We are leaders, ready to face the new and unknown.

To Sum Up

I had so many questions when I left, the main one was: What will this journey do to me? Here are my short answers to the questions I asked then.

- Will our energy and consciousness be raised? Yes
- What is it like to live with very few material things? Easy
- What will our trust and creativity bring? Wholeness
- How can we contribute with our experience? Sharing the transformation

- How does it work to make big life changes, at our age? Great
- What it is like to be a world/ocean citizen? Still new

It feels accurate to declare this journey complete. Something to celebrate! Other things, like my writing and Arthur's guitar playing that started when we were on the other side of the world, are still in their infancy. Like everything else, it is cyclical. By the way, looking for what can be completed is a good way to release more energy.

How do you do to make something complete?

69

LETTING GO IS ESSENTIAL

"Let go of who you think you should be in order to be who you are. Be imperfect and have compassion for yourself. Connection is the result of authenticity." – Brené Brown

We have let go of so much. By leaving everything behind, we showed our willingness to do what was needed. It started with the materials and ended with the mind. The weight loss came as a surprise. It helps of course to have each other.

Looking back, the emotional hardest part was letting go of all our stuff in Stockholm. I can still remember how heavy I felt some days at my office, looking at several boxes of books. Gosh, how hard it was to let them go. Sailors who keep their homes do not have to go through all that. Now, I am very happy we did. It got easier, and now I am eager to let go of what no longer serves me. It is true that less is more.

Someone asked me if I miss anything. Yes, I missed my bathtub, before I found out I had the whole ocean to bathe in. My wedding ring and necklace have been in safekeeping – now, when I am home, I wear them again. I think longing is good, I get grateful the day it is possible again. Like getting flowers on the table!

The trickier parts to let go are the mental habits, like having expectations and wanting to be in control. But even these habits can be let go of. The ocean has taught us to surrender and trust in existence. As adults, we can practice and choose trust, instead of control.

It Starts Early

In a way it is surprising that it is so hard to let go. During our lives we get plenty of practice, even if we do not choose it consciously. It starts when we are babies – we let go of the womb, the umbilical cord, the milk from our mother's breast, then the stroller and the first teeth. We outgrow beds, clothes and interests. We move away from home, change living environments, change partners, change furniture and maybe even profession. And all the time we lose hair and skin, even our cells are replaced. Every second we let go of one breath, to take a new one. We are forced to accept the course of life.

At the same time, we build up desires, attachments and addictions to various things, behaviours, ideas and relationships.

It Is Natural

Everything in nature is cyclical. Letting go and having trust are key functions.

Imagine a tree that does not lose its leaves in the fall, or its flower petals in the spring. It just does not work. For a tree, letting go is natural – its entire system is organised so that new buds and later fruits enter the void. When the clouds are heavy enough, they let go of the rain. We also have to accept what is and trust the process.

So it is with emotions, if I accept what I feel, the emotion will dissolve by itself after a while. If I keep it, resistance will build up. If I accept, receive the message and the gift, everything is so much easier.

We humans have the ability to create new things all the time. Provided we dare to let go and create space for it.

Now it feels like it was a journey in itself to move aboard to live on a yacht full-time. On top of that, we sailed around the world. I am so grateful that we made the brave decision and, as I said, back in Stockholm we felt it was right to let go of as much as we could.

Now I can see that letting go is the key, and that there are so much more than material things to let go of. To make space on a bookshelf did not directly clear my mind. The process is getting easier and easier.

What have you let go of?
What else can you let go of?

70

INVITING THE NEW

"Because you are alive, everything is possible." – Thich Nhat Hanh

Letting go again, and again, gives me space to receive the new. Exploring the hitherto unknown is a powerful way to be alive and vital. To be surprised!

For five years we have been in new surroundings. In New Zealand, everything was the opposite of what I was used to. The water flowed in the other direction, the wind from the south was cold instead of warm, it was summer in December.

Entering something new is a certain frequency. Every new step, every breath, every eye contact. New and fresh. Every person we met was unknown at first. In every new place, we had to find our way to the daily basics. We had to find our way to everything. Even the brain had to create new patterns.

Again, our motto, wave by wave, serves us very well. It means being present and in motion. It is an opening towards something, a direction. Taking the next small step is perfect. That way everything becomes easy, even what at first seems very difficult.

The new has been coming in many forms.

- As gifts in the constant changes and challenges
- as new habits
- as new friends and help

- in learning new skills
- as creativity
- as messages from the soul.

Loving What Is

I have learned to love what is. As sailors, we are very dependent on the weather. In order not to become a victim, I have learned to quickly like and even enjoy the situation.

I stop and look for the opportunity instead. Go with it, is the mantra. There is always an opening somewhere. If possible, we want to go with the flow and have the wind at our backs.

Challenges often come as surprises. It is not something we set up in advance to handle and check off. Who wanted COVID-19? Yet it came and hit us all. The same with the weather – cyclones and tsunamis are not to be trifled with.

In hindsight, I can see that the toughest situations have been the most memorable. They have shown me an inner strength that I was not aware of before. So, it is worth remembering that fear, uncertainty and in my case, the fall were important steps along the way. When it feels the worst, it often marks a turning point.

Being with what is, is the thing.

New Habits

This way of sailing and living on a yacht is very different from the life we had in Stockholm. Then we both were working full-time. This time is now used for creativity and some coaching. Our new way of eating is another big change.

New Friends and Help

We meet new people who welcome us and are kind and helpful. We have new neighbours in every marina. Now and then I have missed my friends. Then it doesn't take long before I meet an interesting new woman to talk to.

I also get the perfect help all the time. Help comes from everywhere; from people, from nature, when the sun rises or dolphins swim toward us. Even the higher spirit protects us. The waves of the ocean carry us, and we have received holy water all the way from the Ganges. We help each other by listening, especially to the intuition.

What helps is to ask. Letting existence know that I would like help and guidance.

Learning New Skills

We also had to learn new things like sailing on the ocean and at night. Diving, front crawl, languages, writing, playing guitar and cooking are other new things we learn.

We have space to learn things that takes a lot of time to master.

Thanks to the internet, we can take part in the most prominent thinkers and masters all over the world. There is great wisdom in nature that we absorb. We are open, curious, willing and explore the new. Especially what is on the edges.

The Learning Process

Being a beginner for so long has humbled me for what it takes to learn something new. As an adult, it can be difficult to accept the stumbling steps and the falls. The so-called mistakes give me information that I am on the way, but not there yet. Being unsure and vulnerable is part of the learning process.

I ask myself if I have hit rock bottom yet. There is a turning point. I have had enough. I have learned the lesson. I see the pattern. I am ready to try something new. I am open to suggestions from others who have been there before me.

A beautiful part of being a beginner is the feeling of wonder. Like a child looking around with wide and open eyes. Amazed by everything.

The learning process is essential for all development. Let us support all beginners. That is when we are most vulnerable and need the most help.

Creativity

We have plenty of time to be creative. What I write comes to me. I wait for it, and sentence by sentence it comes. Arthur learns to play the songs that he loved as a young man. The songs that touch him. We do things that make us happy.

Being curious and welcoming the new is like opening new doors and gaining more access to ourselves and what existence has to offer.

What are you curious about?
What makes you happy?

71

LISTENING TO THE SOUL

"The next message you need is always right where you are." – Ram Dass

It is easy to overcrowd a day. Being so busy with external activities, that there is no time to experience and respond to what is happening within and around me. I cannot hear my soul in all the noise.

In the same way, it can be easy to clog up every little bit of free space. I clogged up my systems in fear of what the emptiness would bring. Like any addiction would do.

It is worth waiting for the soul. The most perfect messages for me will come that way. In line with my uniqueness and path. The inner voice cannot be forced to speak. Nor can it be rushed. We must be still and wait, with patience. Ram Dass also says,

"The quieter you become, the more you can hear."

When I am open and waiting, the next step will come by itself. I do not need to figure out and control everything. The messages that come to me are for me to act on. To trust.

The soul seems to have a deep knowledge about what is enough and perfect. Long lists are not the way to go. The soul speaks to me when the time is right.

Being Still and Wait

When this new idea came to us, it meant a lot of action at high speed. After a lot of movement, I longed for stillness. The same variation as in nature. Different paces or speeds give us different information. High speed is useful when we aim for scanning and quantity. Getting a lot done, like at the beginning of this journey. Slowness helps us go deep, integrate and see the big picture.

This journey has made me slower, more patient and more reflective. So much is unpredictable, and I have learned to have a lot of slack in my calendar. I give projects a longer time and trust that they will be completed in perfect time. I have no deadlines, instead, I focus on the next step.

Waiting, I have learned, is an important part of life. Like the seed, I and my creative babies/projects need time to process and grow. Things are happening, even if I do not see any visual results immediately. I start early to reduce stress and to let the soul speak.

In stillness, a lot happens. I integrate and reflect on what I have been through. I listen to my soul and my body. The body repairs itself when it is not busy digesting food.

The Intuition

The soul speaks to me as intuition. We also call it inklings. Others say gut feeling or hunches. Some say we only have five seconds to grasp them. After that, the mind decides it is not worth trying.

How can I say it is a hunch and not the mind? It is subtle and fleeting. Whispering. Like a brief flash. It always turns out to be right, if for some reason I do not act on it. Then I remember that I had already received a message.

It is a paradox to slow down and act quickly. When we follow the intuitive messages, directly formulated for us, we will have the space needed to act on them as well.

Having connection with our soul is our birthright. The soul is there for us all the time. So is the body when it gets treated with respect.

What is your soul telling you?

72

THE GIFTS

"Each person has a special gift or talent – his individual song of the soul that is unique to him or her alone, which, when engaged in, gives a special and absorbing joy and raises that person's Life Energy to the highest." – John Diamond

What changed, after all the possible releases and new ways of being? A lot – I see them as gifts. The new minimalist lifestyle has helped me expand what it is to be human.

I have gained greater self-confidence. We managed to sail around the world, often in rough weather. Both the relationship and the yacht are intact.

Creativity increased – so far, I have written and published eight books. I have advanced my knowledge and eye for photography. I have made my dream of making picture books a reality. I have published several articles on Noonsite, the Swedish Ocean Sailing Club and Medium.

A higher energy – I feel much more joy, lightness and ease now.

A deeper relationship, both with myself and with Arthur. We both know and trust that we can handle difficult situations together.

A stronger intuition and contact with my inner self, my soul.

My left eye has opened.

My own experience of how big, beautiful and magical our world is.

A body that serves me rather than sucks.

The trust of belonging and being cared for.

More in touch with my feelings, my heart and my vulnerability.
Humbleness.
A greater sense and care for both the details and the big picture.
Wisdom and presence.
Gratefulness.

What are the gifts in your life?

73

WHO AM I NOW?

"As a body everyone is single, as a soul never." – Herman Hesse

Nothing will ever be the same as before. I have not only had many new and profound experiences; now I value my being on a much deeper level.

I was on the verge of staying in Stockholm, which is well-known to me. I am so glad that life made me fall and stop. Something else called me, something completely new, where I have been able to use other dimensions of myself. As I learned in Bermuda, we have everything we need within us. It is true. Life has so much to offer, if we open up to it.

Transition

During the years we have been sailing, I approached retirement age. Like many others, I see this as a time of transition to something new. It will involve creativity and writing. I trust that the rest will develop. I do not need to know now.

The Spiritual Dimension

> "Twice born
> – when you are reborn with your soul"
> – Premartha & Svarup

A course I often think of is the one about the seven-year cycles of life, led by therapists Premartha and Svarup. According to them, each seven-year period in life has a specific life task to solve. The older we get, the more it is about letting go of what we have built up and worked for on the outside. In favour of spiritual development, our wisdom. A way to be reborn again.

I enjoy getting older. It feels like I am getting more and more in touch with the spiritual dimension. When I listen and be still, I get in touch with the universal energy. An incredible opening and expansion. The perfect comes to me. I experience lightness and a sense of flow.

We say that we listen inward to our intuition. At the same time, it gives me contact with the existential – something much bigger than myself. What Jung calls the collective consciousness. It is easy to make wise decisions when I have access to this dimension.

Being Independent

I have been thinking a lot about the question of who I am. There is a course where you ask yourself that question over and over again, for three days. You can imagine how the answers run out, and you become quieter and more amazed. At least, it did for me. I am all and nothing.

Being independent means that I do not belong anywhere in particular anymore. I am. I belong where I am. When I let go of something that I feel dependent on, my ability to be open and present to the moment increases. Something to remember.

Who are you now?

74

NEXT STEP

"Life is not a journey with destination, it's a dance!" – Alan Watts

Our life's journey will continue in a different form from now on. The big moves are over. Now we will take smaller steps. I trust that the next step will come to me when I am still, present and waiting.

The Blessed Islands

My home is where I am, even though I am still a Swedish citizen. We have chosen Macaronesia as our area to stay and explore. It consists of the Atlantic islands of the Azores, Madeira, the Canary Islands and Cape Verde. Nicknamed the Blessed Islands, or the Islands of Fortune. In the Canary Islands, it is 18–25° all the year around. Perfect for Scandinavians like us.

We have long said that we will explore this area to see if there is a certain place where we will stay longer. We already feel that Las Palmas could be the place where we will one day move ashore.

We are curious and love to sniff around to find the new and special from here. They grow a lot of olives, citrus fruits, almonds and small bananas. They have goats and sheep and make delicious cheeses. The sea is rich in fish.

The Sacred Space

Aboard *Vista* is our sacred space. We keep her clean and give her flowers to beautify the inner space. Here is where Arthur and I share what is dear to our hearts right now. We listen and mirror each other. We continue the process of evolving together, to connect on a deeper level. And we unwind our wounded beings.

This is also where we are with ourselves. I listen for my next step and follow my inner voice. Getting reminded of who I am. I do my best to follow my resonance, my yes and light.

Our bodies are part of the sacred space. Keeping our health at its peak with natural and healthy whole foods is important.

Simplicity

We love this simple and quality-filled way of life. It gives us plenty of time to deepen our interests and develop our creativity.

Our dream was to be live aboard and sail around the world. What we got was much more – a coming home, an experience of the inner oceanic freedom.

If you have a dream – go for it! It will lead you home.

What next step is coming to you right now?

OCEANIC FREEDOM

This journey has changed my worldview. I am part of something much bigger. It is even hard to imagine how big. The world was not only what I thought.

I love the Spilhaus World Ocean Map, as it illustrates a new way of seeing the Earth. It highlights the connected oceans instead of the land.

Almost thirty years ago I was given a meditation name, Mukti, which means freedom. This seed has grown, and one day felt it was time to say hello to its origin.

It is easy to feel one with our existence out on the oceans. I feel free in the huge space with horizons all around, and the firmament above. I had the same feeling of freedom and oneness when I gave Oceanic Aqua Balancing. Like the waves have held and carried us, so did I in this bodywork in warm water. Time and space disappeared, leaving me with a feeling of both giving and receiving. Those experiences have humbled me and made me even more amazed by the magic of living on this planet.

Thank you for reading my story!

THANK YOU

♡

There is so much I am grateful for. A life-changing decision like ours does not come out of the blue. I am grateful to all the teachers and masters who have reminded me of my ability to live my life to the fullest, some of them represented by a quote. And to all who have taught me new skills needed for a journey like this.

For this book, I would like to thank Aida Chamiça, Susie Daleke, Inessa Love, Thomas Orbert, Sumitra Sjöström, Robin Von Schwarz, and Lynne Hodgson for early reading and feedback. Lynne has also been my encouraging editor and is the one who makes sure my English is understandable.

For the cover image, I would like to thank Katya, whom I met in the Azores. She immediately sensed my love for flowers. The cover is a montage, but in the original, I am surrounded by Hydrangeas.

I am grateful to my dear and brave husband Arthur, who has been the captain of this journey. Without you, our dream would not come true. Thank you for believing in me and supporting my creativity.

A big thank you to family and friends who have followed us, who keep in touch and are with us in your hearts wherever we are. It means a lot.

I also have clients who have kept in touch and been coached, all over the world. I love you too, thank you for the trust.

During the journey, we have met countless people who have made our journey fun and memorable. We feel grateful to all the sailors who have gone before us and shared their knowledge and inspiration. Especially the Amel family, the members of OSK and OCC. If they did it, we will do it, we usually say before a tough passage.

Jimmy Cornell has been sailing for decades and shared the safe way to sail around the world. His many books, the most famous of which is *World Cruising Routes,* have a place of honour on the salon bookshelf. I would also like to thank Sue Richards, the long-time editor at Noonsite, for her encouragement to rewrite our passages.

With love to you all from Anna

THE AMEL YACHTS

Amel Yachts are based in La Rochelle, France. The Amel company got started 1964 by monsieur Henri Amel, a dedicated sailor who set safety, comfort and quality very high. Over the years he has drawn many models, even though he got wounded in one eye during the war and lost his sight. When he died 2005, he gave the company to his employees.

The great yacht building tradition continues at Amel – the AMEL 50 got voted European Yacht of The Year in the Luxury Cruiser category in 2018. Read more about the Amel story and yachts.

Amel Super Maramu Redline 2000

Our model, Amel Super Maramu Redline 2000, was the last model monsieur Amel himself drew and built for the US-market. It is built for six people, but it is very common that a couple sails her on their own. The American review Cruising World voted it as Best yacht of the Year (2000).

It is a ketch that exists in 479 exemplars, ours is number 435, build 2004.

Facts

15.97 meters long/53 foot, 4.60 beam, 2.05 draft, 16 tons net.
1000 litre water tank, 600 litre diesel tank.
Yanmar 100 turbo diesel motor, bow thruster, generator, solar-panels,

autopilots, water-maker, diving compressor, washing machine, a/c, freezer, fridges and a small dishwasher.

All water goes to one bilge, so less holes through the hull.

The yacht also has waterproof compartments in the bow, stern and engine room.

The deck is made of composite, which lasts better than teak in sun and salt.

A foresail profile and special poles allow us to sail with two flared sails at the same time, "wing on wing", and pull them together on one roller when the wind increases.

ROUTES AND DISTANCES

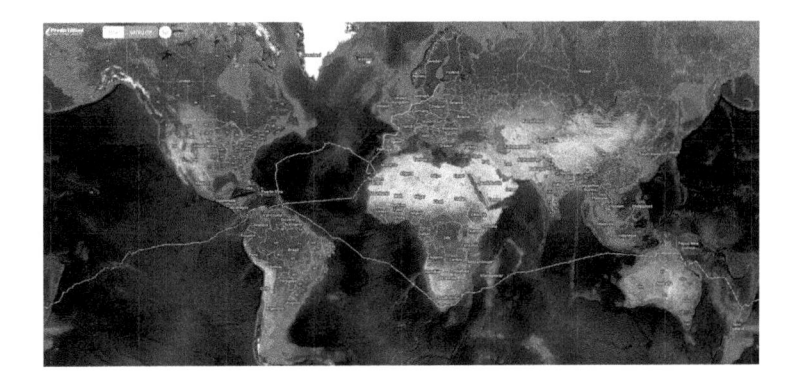

We left Stockholm in May 2019 and returned in November 2024 to the Canary Islands.

The 39,000 nautical mile journey passed through four continents, four oceans and thirty-four countries and islands.

We crossed the equator outside the Galápagos and Brazil. We crossed our track again at Martinique.

Oceans: North Atlantic, South Pacific, Indian Ocean, South Atlantic, and North Atlantic again.

Seas: Baltic Sea, North Sea, Caribbean Sea, Coral Sea, Arafura and Timor Sea.

Continents: Europe, Central America, Oceania and Africa.

Countries and Islands: Sweden, Denmark, Germany, Netherlands, United Kingdom, France, Spain, Portugal, Madeira, Cape Verde, Martinique,

Bonaire, Curaçao, Panama, French Polynesia, New Zealand, New Caledonia, Australia, Lizard Island, Christmas Island, Cocos Keeling, Rodrigues, Mauritius, Reunion, South Africa, St Helena, Barbados, Grenadines, Martinique, St Martin, Bermudas, Azores, Porto Santo, Canary Islands.

Distances

Toulon – Stockholm, 3/7-15/8-18: 3050 NM

Stockholm – Las Palmas, 25/5-25/11-19: 3522 NM

North Atlantic: Las Palmas – Cape Verde – Martinique, 21/12-15/1-20: 2995 NM

Martinique – Panama, 30/1-8/3-20: 1267 NM

Pacific Ocean: Panama – Marquesas, 17/3-18/4-20: 4030 NM

Bora bora – New Zealand, 17/11-7/12-20: 2591 NM

Stockholm – Opua: 16 086 NM

New Zealand – New Caledonia, 12/5-19/5-23: 1049 NM

New Caledonia – Cairns, 30/5-8/6-23: 1537 NM

Cairns – Lizard Island – Seisia – Darwin, 15/5-3/7-23: 1358 NM

Darwin – Christmas Island – Cocos Keeling, 20/7-9/8-23: 2082 NM

Cocos – Rodrigues – Mauritius – Reunion, 20/8-25/9-23: 2511 NM

Reunion – Richards Bay, 14-22/10-23: 1428 NM

Indian Ocean: Darwin to Richards Bay, 20/7-22/10-23: 6021 NM

Richards Bay – Cape town, 26/11-6/1-24: 986 NM

Cape Town – St Helena, 16/1-29/1-24: 1733 NM

St Helena – Barbados, 4/2-6/3-24: 3794 NM

South Atlantic: Cape Town – Barbados, 5527 NM

Caribbean: Barbados – Grenadines – Martinique – St Martin – Bermudas, 14/3-15/5-24: 1520 NM

North Atlantic: Bermudas – Azores, 28/5-11/6-24: 1870 NM

Azores – Porto Santo, 4-6/9-24: 513 NM

Porto Santo – Graciosa – Lanzarote – Fuerte Ventura – Las Palmas, 3/10-26/11-24: 497 NM

NZ, Marsden Cove – Las Palmas: 21 526 NM

ABOUT THE AUTHOR

Anna Eriksson is an Independent Coach and Author from Stockholm, Sweden, with over thirty years of coaching and spiritual development. She has been part of the growing coaching industry in Sweden, primarily through ICF, the International Coaching Federation – as a member and certified coach, chapter host and finally on the Ethics Council.

Anna is the owner of Avalona – Executive & Teamcoaching and she coaches CEOs and senior leaders to become more conscious and transform their inner challenges.

She has a Bachelor of Social Sciences in Behavioural Research, and a background as a researcher at Stockholm University.

Anna also writes for *Medium*, *GROW Magazine* and *Noonsite*.

ALSO BY ANNA ERIKSSON

Leading from Joy

– How to Transform 9 Inner Challenges.

This is a step-by-step book about how you can transform the most common inner leadership challenges and raise your consciousness. You will get sixty practices and eleven real CEO cases. Based on academic and new science, as well as twenty-five years of professional coaching.

Reflections and Photos from the Other Side of the Earth.

– A Play Project by a Sailor and Coach from Sweden.

This series is published one season at a time, sharing a daily photo and reflection, during an unexpected extra year in New Zealand. It starts with *Winter* and ends with *Autumn*.

Messages from Nature

This is a poetic celebration of the beauty and wisdom of Nature. The book includes photos from the dawn and twilight, during a sailing from Nouméa, New Caledonia, through the Coral Sea and around northern Australia to Darwin. While Anna was watching she received messages of wisdom from the nature, connected to what she was experiencing.

Beauty Everywhere

– The Joy of Flowers and Inner Growth

Here you find Annas story with flowers – how they have influenced her life and taught her more about herself and others' unique contributions to this world. It includes photos of bouquets through life.

Anna Eriksson's books at Amazon